Tales from the Trail

Adventure Stories

Big Game Hunting in Northern BC

Dawson Smith

 FriesenPress

Suite 300 - 990 Fort St
Victoria, BC, V8V 3K2
Canada

www.friesenpress.com

Copyright © 2020 by Dawson Smith
First Edition — 2020

All rights reserved.

No part of this publication may be reproduced in any form, or by any means, electronic or mechanical, including photocopying, recording, or any information browsing, storage, or retrieval system, without permission in writing from FriesenPress.

ISBN

978-1-5255-7726-0 (Hardcover)
978-1-5255-7727-7 (Paperback)
978-1-5255-7728-4 (eBook)

1. Sports & Recreation, Hunting

Distributed to the trade by The Ingram Book Company

Dedicated to

Fay who always says I can

Ken who lit the fire and

Denis who fanned the flames

Table of Contents

September Sounds	3
Redemption at Windy Ridge	17
Way Too Close	29
Hard Bargain Valley	37
By the Skin of my Teeth	45
Is Today the Day?	57
Tales from the Trail Photo Album #1	*69*
Pronghorns in the Desert	89
"Odocoileus hemionus" on Horseback	102
A Conspiracy of Events	113
Monarchs of the Peaks	123
Season of the Wolf	133
Cats-n-Dogs	143
Tales from the Trail Photo Album #2	*151*
The Four Year Quest	173
A Northern Adventure	183
"Ursus Horribilis" *The Second Time*	*200*
The Mighty Moose Hunter	211
Of a Quest for Bulls	217
About the Author	231

www.bndproductions.ca

"It isn't the mountain ahead to climb that wears you out, it is the pebble in your shoe"

"Live everyday as if it were your last because someday you're going to be right."

Muhammad Ali

September Sounds

A Big Game Adventure

Sometimes, the difference between having a successful hunt or not boils down to pure luck – being at the right spot at the right time. But luck doesn't always come exactly as expected. The truth isn't always what you want to hear or see, but luck – good or bad – is still luck. That's not to say that local knowledge, research, and or some good old fashion hunting skills don't factor in, they do. However, first and foremost it's all about being in place "A" while your chosen quarry is in place "B". If you are standing a hundred yards to the left, right, back or forward perhaps you won't see what you are looking for or perhaps you won't be able to get a shot away. Sometimes though, long pre hunt planning sessions pay off in spades.

Mid-September, my partner Bill Cash and I pointed his truck north west from our hometown of Prince George British Columbia to a favourite hunting ground of ours with another hunter, Dale Hansen, and hopefully a good helping of luck, in tow. The three of us were heading for a series of basins far off the beaten track in hopes of finding grizzly bears, mountain goats, stone sheep and bull-moose. Bill and I were a couple of the lucky grizzly bear tag winners in the provincial Limited entry Hunting draw while all three of us carried mountain goat, sheep and moose tags in our license booklets.

Usually one notices the colours of fall; golden leafed aspens shaking on distant hills, bright coloured berry bushes carrying all colours of the rainbow, snow white capped mountaintops, purple and blue marbled peaks evolving from late afternoon shadows, silver glistening water ways and those ever impressive fall sunsets. But how many of us ever key on the sounds? The noises emitting from the daily business of the natural world? The audible side of nature?

Does a falling tree make a noise if no one is there to hear it?

Flip flop...

Caw, krock, aack, cacaw, a crow called in the brisk morning air.

"See anything?" my partner asked in a whisper.

"No, shhhhhhhhh" I replied

Aack, caw.

Flip flop...CRACK....flip flop...WHISP...

I stepped ahead a few feet silently cursing the twig I stepped on and the noisy raincoat I was wearing.

"Over here" I whispered to Bill, pointing to the ground.

Stained rust red on the yellowy green moss was a blood trail leading further down the hill. Up rooted clumps of moss marked the way, the exposed decaying forest material lent a musty aroma to the milieu.

"Whatever the bear had dragged must have been heavy" I thought.

Caw..caw..

Flip flop, flip flop...

humph wa urggg – humph wa urggg

HUMPH WA URGGGG, "What's that noise?" I said aloud.

Bill brushed past me and stepped through the willows. Small pecker pole pines stood statuesque in the early morning light, a gentle wind brushed their tops, an errant creak from a tree echoed through the forest. The murder of birds, usually and up till then, raucous, noisy and heckling, grew silent.

Humph waaaa, Ughhaaa hhah uhhaaaa uhauhauha.

"What was that?" Bill turned back and whispered.

Again the sound drifted from the reaches below.

Humph waaaa, Ughhaaa hhah uhhaaaa uhauhauha

Incredibly loud this time. Incredibly intimidating. Our eyes' met.

"Whoa" Bill said. "that's a bear, get behind a tree".

Bill stepped to his right and I stepped to my left.

GGGGGgGggggggrrrrrr.Uphwaa...Gggggggrrrhmp the bear bellowed in the forest below.

The sound resonated throughout the forest, seemingly penetrating the very trees themselves. Bill and I looked at each other. "Oh oh" he mouthed as another roar beat the air.

Then came a pause in the oral ordeal, for perhaps ten seconds or so the bear went silent as it took stock in the situation. Bill and I stood frozen behind our trees. Not even a black fly buzzed in the shadow dappled moody forest. Just when a thought of *"phew that was close"* filtered into my mind a rustle of wind kick started the drama into action again. Starting with a small breath of wind rustling the bushes the noise grew menacingly into a branch snapping, willows shaking growling package of fury and dire. High above a crow swapped ends on the branch it was perched on to get a better view. "Look out Bill its coming" I said, not even bothering to be quiet.

The grizzly was coming up the slope full tilt at us, noisy, belligerent, mad as a hatter. We both fumbled with shaking fingers to load our rifles as fast as we could.

The events that led up to this precarious situation were harmless, almost too good to be true. Bill, Dale and I had been hunting moose, grizzlies, mountain goats and stone sheep on a mountain not far from the swamp where Bill and I walked in on the bear. We had enjoyed a wonderful action packed five-days on top. On the first day, while glassing for sheep, I spotted a bull moose bedded in a high willow thicket. I didn't really want to shoot a moose that far up in the mountains but thought it would be fun to get in close and fool around calling it. Sneaking across a rock slope I got above it and grunted;

"Uggggh....Ughhhhh".

Through my binoculars I watched as it casually looked my way, almost dis-interested.

"Maaaaaugh" I tried my best cow moose call. It looked my way with more interest, "Mraaaaugh" I called again, louder and more desperate. That did it, the bull jumped up and started thrashing a willow bush with its antlers. The noise of breaking twigs carried across the thin air to me. "Mraaaaugh" I moaned again, it grunted and started coming towards me. When the big bull cleared the last bush I noticed how huge it really was, its headgear was far bigger than I had first thought but what really got me was its body size. I called again and it responded and kept coming. I sunk in behind a small tree and let it come. While I struggled to get my daypack in place for a rest the bull moved closer and closer. By the time I got all settled behind my rifle the moose was starting to get second thoughts and had stopped, surveying his domain for the cow. At about twenty five yards away it got cold feet and turned sideways to leave. One shot from my 7mm STW put it on the ground.

I headed back down and got Bill and Dale to help with the arduous task of getting the big bull to camp. Night fell over the high alpine bowl while we attended to the task. Hours later, back at camp Dale told me about the nice billie goat they had seen while I was shooting the moose. They had spotted it high on a grassy knoll and set up after it. Halfway to the goat they peaked over a small rise in the land and came face to face with a grizzly quietly feeding on berries. Bill and Dale moved up a small drainage channel to get a better look at the bear. They poked their heads over a rise in the land about twenty yards from it. Dale got worried about being so close and jacked a round into his rifle. The grizzly spun around at the sound like it had been hit by lightning. Our small campfire crackled to life from a wayward gust of wind as Bill told me that he thought it was going to charge them. "Luckily" he said, "it bluffed towards us a step or two then ran uphill". It was a large bear so Bill was watching it closely trying to judge it as it moved, "then it stopped and squatted to pee so I knew it was a sow" he said. Sitting down on the hillside they enjoyed a feast of blue berries while they watched it slowly march for distant berry pastures.

They had wasted too much time with the bear so decided to descend back to camp leaving the goat for the next day. "That was the closest I've ever been to a grizzly," Dale said from across the fire pit.

The next morning I headed up the drainage where I had shot the moose while Bill and Dale headed back to see if they could find the goat. The billie was in the same place and they made a roundabout three hour circular stalk to get above it. Once in position they watched to make sure it was a billie and not a lone nanny. Meeting back in camp that night Dale replayed the hunt over and over while a glorious treat of goat steaks fried over the fire. I had spotted quite a few sheep in the distance that day so it was decided that the next morning Dale and I would head out for a closer look while Bill would poke his nose further up the basin where they had shot the goat.

Dale and I started spotting sheep almost the second we crested the ridge that hid the small box basin.

Snow-capped serrated peaks walled the bowl on the north side while long talus and grass slopes led high away on the south side. We had just located a couple sheep lying on a slope when I spotted a white spot high on a rock bench across the basin. Swinging my spotting scope around I zoomed in on a goat just as it stood up to stretch.

"Goat" I whispered in the rustling wind. Dale turned around and trained his binoculars on it. For the next hour or so we watched it closely, finally determining that it was a mature billie. Down below a fog bank casually flowed into the basin; like a tidal wave it pulsed in and out, one minute obscuring our vision then clearing for a second or two, finally it wrapped the entire basin in its misty embrace. We took the opportunity and moved across the valley floor towards the goat. When the ground fog cleared we were hunkered down at the bottom of a rockslide a couple hundred yards below the billie. Up above, the goat sensed something amiss in its solitude and stood staring straight down at where we hid. It was a stalemate, we couldn't get any closer and the goat just stood looking over its desolate realm. Satisfied that nothing was out of place it finally turned around and lay down. I jumped at the chance and crab walked into a small runoff channel that offered the best cover I had at hand to get higher. "You better wait here" I whispered to Dale. Once in the depression I was able to sneak straight up until the goat was fifty yards opposite. I couldn't see it but knew it was bedded just over the edge. Stealing a quick look around I stepped over to a small bush I could get behind for cover when I kicked a rock loose;

clack..clatter..clack....shhhwec it made clattering down the hillside.

The goat stood up and walked over to investigate the noise, spotting me he turned to depart to higher ground. In that instant I could see that he was a dandy. I sat down and found him in my scope. Just before he disappeared over the rock I fired.

"Three for three" Bill said as we sat around the campfire that night eating more goat steaks. High above a myriad of twinkling stars added their primordial presence to our good fortunes. Dark unblemished blue sky's promised another day of magic to an already magical trip. However sometime in the night a brisk wind whipped the nylon tent fly and raindrops fell like quarters from a slot machine. The next morning a wet dankness filled the high mountain air. Undeterred Dale and I headed for the distant sheep slopes. By noon we had spotted numerous ewe and lamb groups but no rams. Bill was hiking in an adjoining basin, he had seen sheep there as well but also failed to find any rams. In the late afternoon I climbed the basin wall to peer into the next bowl. After a sweat dripping three-hour climb I was met at the top by thick fog, snow and wind.

"Perhaps we had used up our helping of good luck the previous few days" I thought

Back at camp the weather continued its downward spiral; a heavy thundering rain was pushed by gusting winds, snow squalls replaced the rain as the temperature dropped further at nightfall. The next day we hiked into a long basin but by noon had had enough of the wind and snow and retreated back to camp. The snow line slowly lowering down the slopes above us lent to our decision to get off the mountain while we could. The trip off the mountain the next day was full of adventure, hard work, sweat and frustration. By nightfall we were still miles from the pickup and a rainsquall had gripped the lower valley. The next morning found us pitching camp by a river miles from the mountain. We were all soaked from the night before so erected drying racks around a fire and set about getting camp together and drying our gear. My boots were soaked so I hammered a couple sticks into the ground by the fire and hung them upside down to dry. I had brought a pair of beach sandals along as lightweight camp wear so "flip flopped" around camp tending to my chores. We had just got the tent pitched when a tempest of crow squabbling drifted to our ears from a swamp a half mile or so away.

"Let go see what they are all up in arms about" Bill said, "I bet something's got a kill down there".

And that started the hair-raising episode at the beginning of this story. We were just going to go saunter down through the forest to the swamp to see what the birds were making such a fuss about. Me in my beach sandals, Bill and Dale in soggen clothes, all of us carrying empty rifles. I guess we knew that most likely a bear had a kill but most of the time when the birds are noisy that means their feeding and for that to be true then the bear, if there indeed was a bear, wasn't at the kill.

Hmmmmm.....so much for that bit of wisdom.

At the edge of the forest, pile upon pile of bear scat marked the ground. That should have told us what was what but we decided to plod ahead anyhow, bravado in numbers. Who's scared, not me, how about you. Dale bailed out on us once we got fifty yards or so into the forest. "I'll go back up to the truck and wait" he said. A small patch of willows lay in our path. Bill took one side and I took the other, the bear had dragged its bounty straight through the middle of it. At the end of the willow thicket I walked the length of a blown down tree and jumped off at the end breaking a stick in doing so. "Crack". Stepping ahead a few feet or so a hungry branch reached out snagged my raincoat, straining from me stepping ahead it bowed and finally let go, "Whisp".

Bill whispered "see anything?"

Unbeknownst to us at the time the grizzly had buried it's kill about fifty yards farther down the slope from where I jumped off the log. It on the other hand was bedded on a natural ledge about twenty-five yards off to the right of the pile. I can only assume that it had listened intently as we flip flopped our way the two hundred and fifty or so yards down from where the truck was parked. One can only assume as well that there was a boundary that decided its course of action. One minute it lay listening to us content that we were far enough away from its dinner, the next it was pure business.

Kind of like a line in the sand *"I dare ya to cross it"*. The only problem was that we weren't looking for a line in the sand.

GGGGGgGgggggggrrrrrr.Uphwaa...Gggggggrrrhmp the bear bellowed in the forest below us. Trees shook, branches broke and the ground seemed to bounce. The noise level was incredible.

"Look out Bill its coming" I said, not even bothering to be quite. The bear was coming up the slope full tilt at us, noisy, belligerent, mad as a hatter.....

We both fumbled with shaking fingers to load our rifles. I wasn't aware of where Bill was or what he was doing, but the next thing I remember is stepping out from behind the small tree with my trusty 7mm STW loaded, safety off, and shouldered. With my head cocked away from the stock of my rifle I searched for the bear in the thick brush below me. The hill stepped off flat to the swamp fifty or sixty yards below us and I scoured its brush-choked expanse for the bear. The noise grew louder and louder as the bear worked itself closer and closer, madder and madder as it came. "Where is it?" Bill said, "I can't see anything". I vividly remember starting to turn to answer him when a dark *"object"* moved in the lower part of my glasses. Looking down I saw that it was the grizzly. He had stopped for a second and stepped out from between two tree's looking straight at me. Instead of being fifty or sixty yards below he was standing there a scant twenty feet away. It saw me move, I saw it move, I swung my rifle settled the crosshairs and pulled the trigger.

Kaboommm…. I chambered another round, Kaboomm……Bill was yelling "where is it, where is it?"

The bear fell from my sight, bushes shook as it made its way down the slope away from us roaring as it went. "Bill let's get the hell out if here" I yelled. Shoulder to shoulder we made our way back to the pickup. "I think I hit it good" I told Bill. "Let's go back to camp and give it time to expire," I said. "I'd really like to put my boots on before going back down there."

Back at camp we brewed a coffee and replayed the event. Bill hadn't seen the bear, like me he was looking down on the willow flat for the bear. Dale told us that he could hear the bear clear up to the truck,

"I heard it growling and the bushes breaking" he said "that was a hairy noise way up on the road."

"Phewwyyyy" I said, "it was even hairier in the bush."

After an hour or so we decided to go back down and find the bear. Another hunter in the area had told us that a few days earlier he had seen a sow with three full-grown cubs around the swamp where I had shot the bear. We didn't know if I had shot a lone bear, the sow or even perhaps one of the grown cubs. The task at hand wasn't on top of our "fun things to do list" I can tell you that.

Back at the spot we loaded our rifles, chambered one on top, something neither us ever do, and waded into the forest. Every creak or moan from a tree or bird noise had us stopping and looking, fully expecting fury and evil to emerge with murderous intentions out of the forest gloom. But it was anti climatic, once at the tree from where I had shot, Bill spotted the grizzly piled up in the willows below. I showed him where it was when I had pulled the trigger, six paces! From his tree he wouldn't have been able to see it until it was right on top of him, there were too many bushes and trees in his way. With senses on full alert we tip toed down to the fallen bear. It was dead and most likely had been when it rolled off the hill. My first shot had broken both shoulders and was all that was needed. It was a large cinnamon blond boar. Measuring 7'2" it carried a massive frame, huge teeth and those ever intimidating front claws. "Phewy" I said to Bill, "that was as spooky as it comes." Bill didn't even answer, for him that meant he agreed.

At camp that night another party of hunters came over to see the big boar and swap tales. They had seen another large grizzly a few miles further down the valley that morning. The next few days we hiked up and down the valley looking for that bear but never found it. The days did however bring a small bull moose for Dale and then late another night we spotted a big bull moose high on a ridge. Scrambling to get into a shooting position Bill motioned for me to call.

"Mrauuuuuuuug" I moaned into the still air.

The bull grunted its response and came to my lovesick cow call like a runaway freight train. I kept calling as the moose closed the distance. As long dark shadows chased light from the valley the bull stepped out of the bush ten yards from Bill. "Even you couldn't miss that shot" Dale and I goaded Bill as we quartered the moose in the dark night.

We spent another couple of days in search of the dark boar the other hunters had told us about but never saw him or any other grizzlies for that matter. We did though, spot and call bull-moose every day. "Urhhg" "Urhhg" their pathetic guttural grunts would answer our forlorn cow calls "Mrauuuuugggs." And every day we listened to the hectic bickering of crows, cheerful chattering of feeding dickey birds, and distant quacks of ducks.

I'm now convinced that a falling tree would make noise even when there's no one there to hear it, noise is an integral part of the natural world. Just as sure as I am that the noise the bear had made was an adventure all by itself. It was an unbelievable experience, not one soon to be forgotten and not one any of us are eager to repeat.

Redemption at Windy Ridge

A Sheep Hunting Story

"What happened?" Bill asked.

"Oh that" I said, pointing back towards the slope I had just walked back from. "I was just practicing the ancient art of *Belomancy*."

"Belawhat" he replied, looking at me like I was talking Egyptian or something

"*Belomancy,* pronounced belo-man-cy, it's an astrological occulty magic thing, you read the positions that a handful of randomly shot arrows land in the ground" I said "it's supposed to help tell your future, kind of like reading tea leaves or tarot cards"

"What the heck are you talking about Dawson, did you fall on your head or something. Arrows? Tea leaves? Tarot cards? What does any of that got to do with sheep hunting?" Bill looked at me like I had lost it or maybe had taken a short step off a high rock pile.

"Oh ya, it's kind of cool, all the rage in California these days" I continued.

Bill looked around the high basin we were standing in and took a step back, his eye-brow's raised in concern. The wind had picked up again, leaden skies threatened, off in the distance a small rockslide clattered, "have you gone nuts?" he asked.

Bill and I were about sixty kilometers off of the Cassiar highway in north western British Columbia. We had trailed four horses and a donkey into the area days earlier. A few years earlier we had traveled into the area on horseback and had encountered a never-ending succession of muskeg bogs and mud-holes. On that trip we had both shot exceptional mountain goats and saw some of the most untouched, game rich, awesome country BC has to offer, but at a price. We had limped out a string of worn out, tired and hurt horses, in fact we were lucky to bring the whole pack string out. Never, we both had said, never would we attempt to cover that trail on horseback again, it was too tough. Never say never the old saying goes, we decided to give it another try. The reason for our change of heart was success in the annual Leh draws. My partner Bill Cash had grizzly bear and mountain goat tags in his shirt pocket, while I had a stone sheep Leh in mine, having filled my grizzly tag during a spring trip to the area. The previous four days had been hell, just as we remembered. The poor horses battled through the bogs, traveling was tough and painstaking slow but eventually we gained ground on the distant mountains. Unfortunately we had to leave one set of pack boxes and a saddle along the trail when the donkey powered out. We'd have to figure some way to get the gear out but for now we were heading in. Once we gained some altitude the muskeg was left behind, the going was easier for the horses and we broke through the dense timber into the high valleys and basins that we had reveled in on that trip two years earlier.

My rendition of *'Belomancy'* was in fact nothing at all like the real thing or anything at all to do with divination. We had just spent the entire day stalking a band of eleven rams, in that band there were two awesome sheep, the biggest was a heavy horned, dark bodied, grossly broom'd off old character. Another ram had wide sweeping oak coloured horns curling considerably beyond his nose. While there was a couple more that would pass the full curl test, the two big ones were what we had come the long way from our hometown of Prince George to find. We had spooked them off the mountaintop first thing that morning when we foolhardily crested a ridge astride our horses. As they fled across the open bowl we sat down in the blustery wind and watched where they went. A few hours later we headed after them. Later that afternoon I finally had crawled up a rockslide, peered across the abyss that lay between the bedded rams and myself, picked out the exceptionally wide ram and shot. Then another................... okay, then another.

Then maybe a couple more. I couldn't assess a random landing pattern to see what the future had in store for me as I was shooting a .30-06 not a bow and arrow, and in fact hadn't even seen where the bullets landed.

Bill kept looking at me in a quizzical way during the four-hour hike back to the waiting horses. We were camped low in a valley about an hour's ride from a vantage point we aptly named Windy Ridge. That's where we had busted the sheep off, and that's where we rode out to most days to spot the myriad of valleys, slopes, and mountaintops.

The next day we left the horses in camp and climbed up a closer mountain. We had seen a pretty good size billie up there a few days earlier. During the mornings drizzle we climbed up through thick stunted trees and fought the ever-present willow thickets. Just as we broached the tree line a glorious sun melted off the clouds. In the distance we watched the retreating clouds shed a wet trail across the adjoining mountains. Vivid rainbows replaced the grayness as the sun continued it hurried advance.

"There it is" Bill whispered. Above us on a ledge lay the white monarch. Our path would leave us vulnerable to his watching eye so we opted to head across the face hoping to come down on the goat from above. Climbing into the next basin we saw that a top approach wasn't going to work, the goat was bedded below a series of shear faces. "We'll have to cross across that rock slide," I said to Bill, gesturing off to his right. I decide to head down and circle under the goat while Bill struck off across the rocks. He was going to try and reach a boulder pile a couple hundred yards above the billie, from there he should have a perfect shot. I went down to save making excessive noise crossing the slide. Small rockslides erupted from my feet as I pussy footed off the saddle, Bill vanished in the wall of rock he was crossing. About an hour later

I heard the report from Bills rifle rebound throughout the mountaintops, then another, then silence.

Bill's smiling face-met mine as I followed a small ridgeline to the top "there he is" he said. Looking over I saw the fallen goat. "Perfect stalk" he yelled. The goat had rolled off his perch to come to rest only a few yards from his bed. Bill excitingly told me that he had crept up on the goat undetected, but at the last minute the billie stood up to look around. "He didn't know I was there," Bill said "but he knew something was up." We caped out the mature goat. Loading deboned meat, hide and skull into our packs I led the way down off the mountain.

"Now that's the way to do it Dawson" Bill goaded me as goat steaks sizzled in the fry pan. "Were not done yet" I said, "I'll catch up to those sheep tomorrow".

A brisk northern wind grew to hurricane forces in the night. I thought our small tent was going to get blown off the mountain, then the rain came, then the snow. "Tomorrow eh?" Bill grumped from his sleeping bag. "Ya right." The night passed, bringing the morning. We saddled the horses and rode off, but between the wind, snow, and fog, we could barely see our noses, let alone any sheep. During one clear spell I was spotting down a valley and saw a bear, a big bear, "grizzly" I said to Bill. The bear was feeding on the lee slope of a small mountain perhaps two miles away, we watched it between foggy patches hoping to pin it down so we could take after it. However, the bear didn't co-operate, it fed along the hillside, we watched, then fog would roll in like a tidal body of water. When the fog retreated we would scour the hill for the bear, just find it in our soggy optics and the fog would wash back in. Back and forth our hopes were pulled, in and out, up and down. Eventually we couldn't locate it, the bear must have fed lower in the thick bush. "If it was clear" I groused "we could probably go after that bear". "This is a waste of time" Bill said "lets head back to camp."

Back at camp we spent the remainder of the miserable day locked in our sleeping bags or scrounging for dry fire wood. The next day mirrored that one, the next the same but snowy. Bill finally asked me during one sitting in the tent because of the rain spell.

"What was all that Belawhatchacallit crap the other day?" he asked.

"I blew it" I said "just down right missed a shot that I probably should have made. It was longer than I thought but still, shouldn't have missed. Any ways I was just pulling your leg with all that Belomancy junk."

As the days slipped by my thoughts turned back to the band of sheep, their images flowed through my mind as I lay awake, talk about counting sheep to get to sleep. We hoped the weather would change so we could try our luck on Windy Ridge again.

As if to answer my prayers the skies cleared and I was awakened by the sound of noisy, busy, birds. "Bill" I yelled, "wake up lets go". Coming out of the tent I looked up into a clear baby blue sky, to the east an amber tinge painted the air. It was going to be an excellent day. Bill groggily came to life as I saddled up the horses, "better throw a pack saddle on Luke" he said, "might be a lucky day." The horses were eager from three day's rest, they almost ran the whole way up to the ridge. Leaving them hobbled we set up spotting scopes and scanned the distant hills for sheep. Aside from a band of ewes, the mornings spotting failed to reveal anything else. "Let's head up that valley," Bill said, pointing to the east. Dropping back off the ridgeline we rode into a pristine alpine bowl. Marmots scurried toward their holes squealing high-pitched whistles in displeasure at the invasion. Riding further we stopped and scanned the hill behind us. High in the red rocks lay a ram, under scrutiny of the spotting scope he was an immature ram. Farther up the basin we topped a grassy knoll and sat down to glass. Sheep were everywhere on the emerald green blanket, ewe's and lambs cavorted and fed, the younger ones chasing each other round and round. "There's gotta be some rams around here" we said to each other. But nothing showed. Just as the wind was picking up and darkness laid claim to the high country we made our way down the mountain to camp.

The next morning I was making coffee while Bill tacked up the horses, "let's leave the pack saddle behind today" I said, "change of luck." We rode out that morning with our two riding horses under saddle and the three pack horses bare.

Sitting behind rocks to shield ourselves from the excessive wind we scanned the familiar slope for sheep or bear. "Let's ride over there" Bill said, pointing to the far slopes "we've looked at it so much we might as well ride on it." "Ya" I replied "and we should poke our noses over the far ridge into the basins on the other side."

Leading the horses down the steep shale slopes we gradually crossed the basin, we stopped along the way to glass but the land looked bleak and desolate.

At noon a small snow flurry engulfed the valley as we crossed further. Details on the far slope became distinguishable without binoculars or spotting scopes as we closed in. "Still can't figure out where that bear went to" I said during one stop, "lets head up over there" pointing towards a saddle between two mountains. The snow quit, birds fluttered in the returning sunlight, a small creek led our way, noisily dissecting the basin. One of the packhorses stopped to feed on a small field of green grass by the creek, the others joined in. "Let's have a look with the binoculars," Bill said. "the horses can feed for a bit." We scanned the open slopes, seeing nothing we set to head off.

"Sheep" Bill whispered. Focusing on a snow patch low on the hill I saw a young ram walk out on it, the whiteness was a perfect backdrop. His ¾ curl horns easy to make out, just below the young ram I saw movement. "What's that below him in the dip?" I said to Bill. "Can't see anything," he said, "oh wait, now I see it." "I think it's another sheep" he said. As we stared at the spot a huge ram lifted his head, the snow a perfect canvass. "Not a question" Bill whispered, "that's a great ram." I set up my spotting scope and watched the big ram as he crossed the snow patch as well. His horns easily passed his nose, they swept wide and tall. While not as heavy as the dark horned ram, nor as wide as the younger ram from that last group, the ram was perfect. They were climbing the hill towards a grass-covered bench above them. "Bet they bed down up on that bench" Bill said. "You stay here" I said heading off towards the rams. I had a small drainage to go up, once up on the top I would be able to side hill across to where the rams were. Most of the way I was hidden from their view, I only hoped that they would stop on the bench instead of carrying on. The climb up and crossing the hillside was easy, I had one more boulder-encrusted bench to cross and then thought the rams would be below me. Crawling the last fifty yards or so I peered over the edge, nothing. *"Now what"* I thought. Slowly standing up I saw the back of the big ram's head over the next rise. I dropped down and climbed a little higher up the talus slope, step by step I carefully placed my boots. *"Don't kick a rock"* I warned myself. *"Another twenty yards, get to that black rock"* I thought. Another five yards, there, movement. Just then the ram moved. I kneeled to see better, a rock, set loose by my hasty move, careened noisily down the hillside alerting the sheep. The wind, up 'til then had been light, but an errant gust picked up a small pocket of sand up and carried it across the hillside. Above me a rock tumbled off the face clattering down, clack..clack..ckackkkkkkk..it tumbled. The ram stood nervously, turned towards the noise, changed his mind and turned to look over the valley below. Far below by the creek the horses loomed conspicuously out of place on the alpine floor. The big ram looked around again turning broadside to me. I raised my rifle locked the crosshairs and shot, the ram took off. *"Not again"* my mind screamed. I stood up, from there I could see the ram. He was down on the grassy bench. I rushed over to him, scattering rocks as I descended; the smaller ram ran off a bit and stopped. I touched the big ram, the younger one came closer clicking his curious click. Walking over to the edge I yelled for Bill to bring the horses up.

However, he had a ringside seat and had watched the drama unfold and was already on his way up.

"Betcha wished we had saddled Luke up this morning" he said walking up to the ram smiling like a cheshire cat.

The wind stayed with us as we snapped off a bunch of pictures. After caping out the sheep we made a basket out of my saddle and loaded the hide and meat in it. We had a nine-mile ride back to camp that night, the wind howled, the sky at times blue, at time black, threatened but held it's liquid cargo. The last climb before camp was over the top of Windy Ridge. At the top I paused and looked across to where I had taken the ram. What an awesome turn of events, I had thought that another chance at a good ram wasn't in the cards, redemption never felt so good.

The next day we saddled up the horses threw a packsaddle on Luke and set off for the far bowl where we had seen the ewes. We were going to explore its reaches hopefully finding a grizzly for Bill. As usual the persistent wind gusted in our faces as we made the descent off Windy Ridge. Following the river down I had just looked back to make sure all the horses had come off the slope when a fast movement caught my eye. Bill had leapt from his horse and was frantically pulling his rifle out of the scabbard. Jumping off my horse I rushed forth to see a grizzly crossing the stream. It ran up the far side behind a willow thicket and stopped. Bill ran ahead a few yards and lay down, the bear stalked ahead a few feet confused at its interruption. It swung back and cleared the bushes, I grabbed the two lead horses and watched.

Bill crawled ahead a few feet and settled his rifle, the bear stepped forward and Bill shot.

Later that night at camp we relived the whole trip, snow fell quietly as we fried sheep steaks over the fire. Getting in had certainly been an adventure, the first mess up with the band of sheep threatened to stain a negative patina on the trip. Then the adventure for Bills goat turned the tide back in our favor. The ram I shot gave us momentum and finally conclusion with Bills excellent grizzly bear. Sitting back we marveled in our good luck, I wondered what the pattern of shots I had missed at the sheep would have told us.

"Belomancy" I muttered aloud.

"Oh don't start that crap again" Bill replied, "I'm too tired."

We packed up the next morning and started what would be a marathon twenty-six hour non-stop trek out to the pickup.

Way Too Close

A Cougar Hunting Adventure

"A twitching tail revealed the crouched lions position, an almost inaudible snarl sprang forth from its partially open mouth, ivory white fangs stood as testimony to its savage fierceness."

The great author Jack London, in his novel "Sea Wolf," analogized the never ending struggle between life and death as that of pieces of yeast, every living being in the universe comprising of a single piece of yeast. Living or dying in this fermenting world breaks down to the power of each single piece of yeast. Lessor pieces succumb to greater, survival of the fittest, the big and the strongest eat the weak and the small, to the victor the spoils!

I'm not all that sure that London was referring to modern day people, or if he was referring to early man's struggle against one another, nature's beasts, or even in the context of man against beast at all, but at the moment I saw the crouched cougar I was definitely feeling like the weaker of two fragments of yeast. The big mountain lion had far superior weapons than I. Its teeth and claws easily outclassed my humble physical attributes, its mere physical presence had me shaking with fear. There wasn't a doubt as to what piece of yeast was the greater or strongest. But, in my trembling hands I held the equalizer; the ace of spades, the proverbial card up my sleeve, my trusty rifle!

Unconsciously I raised my rifle; squinted through foggy wet glasses, and further through a fogged scope, found a point of aim and squeezed, no that's not right. I jerked the trigger.

I was still-hunting for cougar on the north east end of Vancouver Island in British Columbia. The valley my partners, Harv Knapp, Stu MacFarlain and I were hunting had long been a haven for the elusive *"Felis Concolor."* The three of us had hunted black-tail deer in the area many times before and had found or saw cougar or cougar spoor on each visit. It was during an early season hunting trip for deer that we had decided to hunt mountain lions when the December season opened.

One day on an earlier weekend trip, Harv and I had come across a mother and her kitten on the valley's floor. Under leaden skies we watched the two cross through an opening ahead of us, they both stopped and trained their golden eyes on us before disappearing into the thick foliage. The next morning I happened upon a lone cougar in the dense forest; I was sneaking along a ridge in the timber, on the next ridge over a mountain lion appeared like an apparition in the veiled fog, although I had scanned the ridge with my binoculars and not seen anything, it had just appeared out of nowhere. I quietly sat down and watched as it prowled through a small clearing in the thick timber. Slowly and methodically it explored the clearing for something to eat, finding nothing it faded away into the fog as magically as it had appeared.

Opening weekend of British Columbia's mountain lion season found the three of us making the five-hour drive from our hometown of Victoria to the valley. A fresh dumping of snow greeted us. This was perfect, for without snow – fresh snow – our chosen method of hunting the elusive animal would have been much harder. But the deep snow also hampered our ability to reach the far end of the valley where we intended to hunt. We had brought motorcycles but they failed to navigate the deep snow. We settled on camping lower in the valley and would concentrate our efforts on the lower timbered slopes.

Early the next morning, over oatmeal and steaming mugs of coffee we made our days plans. Harv and Stu would ride their motorcycles as far up the valley as they could, I would find the closest spot and climb into the towering Douglas fir forest. They would climb into the trees and work back towards me, I would slowly work towards them. We had planned on meeting sometime during the day, hopefully the deep snow had forced the deer down to lower elevation, if the deer moved lower down the mountain then we assumed that the cougars would as well. Each of us would slowly sneak along in the timber hoping to find fresh tracks. From there it would be a simple follow the dotted line approach.

Bidding good lucks to each other we set out in the dark stillness, snowflakes the size of quarters fell from the starlit sky, a steady hiss from the falling snow broke the eerie silence. Finding a likely looking spot I headed across a clear cut for the timber. Thigh deep snow slowed my progress; it was like walking in porridge, down my foot sank until it reached a solid purchase, up I swung my other leg, down, up and so on. I had only gone halfway across the clearing when dawn sent the first golden fingers of light over the adjacent mountaintops. Stopping to gather my breath, I scanned the forest's edge. I was less than halfway and already soaked to the bone, the exertion of crawling through the deep snow also had me sweating and my glasses were fogging over. On I went.

Just before entering the forest I again scanned the edge with my binoculars. Of to my left the pillowy mantel of snow on a blown down tree looked skewed, looking closer I made out tracks leading its length. A thump of excitement came from my heart, one end of the tree was hung up high off the ground. Only a cougar could mar such a precarious surface with its trail, and since it was snowing heavy the tracks must be fresh!

Luck is an integral part of hunting, any hunting venture needs some measure of luck in order to be successful. It was lucky that I chose to cross into the trees at the spot I had, finding tracks so quick in the hunt was an extra-large helping of good luck. But lady luck can be a harsh mistress, if the snow wasn't so deep I may not have seen the tracks as I would have probably already reached the timber before daylight. Lucky or not I grabbed the opportunity and charted a course to connect with the cat's trail.

Slowly I crept into the shadow filled forest, each and every second fully expecting to see the cougar exposed in the knee deep snow, I had dreamt this very scenario. About fifty yards into the forest a small rock outcropping rose before me. A wind dropped tree had fallen from its top and lay lengthwise down the incline, a trickle of water cascaded from the rocky top, water droplets glistened like diamonds in the mornings shine. Quietly I climbed up the incline beside the tree. Peering over the top of the incline I looked deeper into the timber, deep virgin snow blanketed the forest floor. Deep, powdery unmarred snow.

A small fleeting feeling brushed over me, I felt a presence before I saw or heard anything. Somewhere deep in my mind a gong was going off. I was going around the butt end of the tree when the cats twitching tail grabbed every ounce of my attention.

Apparently the mountain lion had watched me climb the rock face, at one point it must have been no further that ten feet above my head. I can only assume that it was curious of what I was. By some divine quirk of fate we had arrived at the same place, at the same time, on the same day. Lady Luck again?

Slowly rotating my head, my eyes followed the nervous tail up the body. Before I realized just what was what I was looking hard and fast into the menacing eyes of the cougar. That's when it actually hit me, it was a mountain lion, "*Felis Concolor*," the same animal R.D. Lawrence called "The Ghost Walker" in his book of the same name.

"Too close" my mind screamed "back up." The tawny beast crouched nine feet away, its eyes boring deep holes in mine. I stood there as if a statue; frozen in time, by the moment, fear, surprise, or all. At the time I didn't know what emotion anchored my body and as I sit here today the same feelings raise goose bumps on my arms. Seconds crept along like hours; my heart was doing its best to pound out of my chest, my legs, hard and strong from years of backpacking, felt like jell-o. To make matters worse, my glasses were slowly fogging over again.

I often wonder what predatory instincts remain in us; sure I shoot at the range allot so my rifle is intimately familiar, and sure I have shot many game animals but usually not at nine feet and certainly not one that was displaying the attitude that this one was. I still do not remember shouldering my rifle or selecting an aim-point or even if I did select an aim-point.

My rifle exploded; the boom defiling the icy stillness, the cat disappeared from the scopes murky field of vision, or I blinked, whatever the reason it was gone from my scopes sight. Lowering the gun I saw the cat lay motionless in the snow inches from my rifles barrel. I grabbed the reins on my shattered emotions and backpedaled fast, chambering another round as I did. However, the first shot proved to be enough, the 140 grain Nosler Ballistic tip had found its mark.

I sat down in the snow, in a more comfortable distance from the fallen cougar and watched. Twice I gathered my nerve to venture close, twice deciding that to wait a little more time would be a good idea.

I wouldn't call myself a green horn when it comes to hunting or in dealing with wild animals, but, approaching that cat as it lay in the snow was one of the scariest things I have done. My fears were without warrant, it was dead and had been a micro-second after I had pulled the trigger on my trusty .270. What a magnificent animal, it was a mature tom, most likely it had been foraging the forest's edge looking for deer. With saucer size paws crowned with razor sharp claws, middle finger thick canine teeth and a sleek powerful frame, a mature mountain lion could be nature's finest predator. I was joyous, awed, the feelings rushed to me as I touched the downed cat. As I sat in the snow beside the cat I replayed the event and my reason for hunting the cougar in my mind. Mountain lions are one of my favourite animal, along with wolves and bears. For me they embody the mystical raw personality of wild animals at their best. Primal to their core they continue to thrive in an ever decreasing, ever demanding, changing wilderness. Cougars live mainly solitary peaceful lives mixed with raw violence in order to live. I harbour no regrets for killing such an awesome animal, and in fact was reveling in the adventure that the hunt had provided.

To slowly hunt through the big timbers is my favourite hunting style, stalking in fresh snow is only icing on the cake. I usually like the tranquility of hunting solo, one with nature. To sneak close to un-molested un-aware wild animals is the thrill I get from hunting, whether I make a kill or not. But on this day, this particular adventure, I would have given my eye-teeth to share it with someone. It was a powerful experience.

Paying respects to the hunting gods I slowly set about dressing the cat, he was huge, it was all I could do to maneuver him into a better position for the task. I dragged him as far as I could and then left to get help.

Harv and Stu showed up a little while later to help me retrieve the tom. We all marveled in the uniqueness of such a hunt. Harv said it was probably a once in a lifetime opportunity, but at nine feet, it was way too close.

Hard Bargain Valley

A Sheep Hunting Adventure

"This is no game for the weak-kneed and fainthearted. Hunter success is not high, not because there aren't enough sheep but because there aren't enough people with the temperament to become sheep hunters."

Jack O'Conner

"Wow" I said, "the third one from the right looks pretty nice Geoff."

"I thinks it's that caramel horned ram we saw a few days ago." He replied.

My partner was set up behind me looking at a mountain goat on the other side of the valley, glancing over my shoulder I saw him swing his spotting scope back to the bunch of stone sheep rams that were milling about on the scree slope above. The two of us were hunkered down in a batch of trees, in a rain squall, at the end of a closed valley. We had gotten there a few hours earlier while the group of rams, thirteen in all, were bedded down high in the rocks. As we watched they were slowly getting up, stretching and hopefully they were going to move lower onto a glacial mound to feed.

"Ya I think that's him," Geoff whispered "that is a nice ram."

It was about one in the afternoon, we had spotted the rams from the end of the valley about five kilometers below at first light. Our trek up the valley that morning had been cold and wet. Misty rain tempests flowed over the crest at the end of the basin and marched steadily down the valley shedding their cargo as they went. Low, wet, clammy fog rolled in and out between the squalls as we hiked up the creek bed in shoulder high willows.

Bill Cash, Geoff Helfrich and I were in northern British Columbia in mid-September. We were hunting a set of valleys for stone sheep. In a novel I had read the summer before a character in the story described her *"paradise"* as a walled garden. The series of valleys we named "Hard Bargain Valley" met that description perfectly. The valleys were narrow but quite long, the walls rose sharply from the creek bottom to become grassy slopes with small willow patches here and there, then upwards to talus slopes before culminating in rocky ramparts. Along the east side there are a succession of short little boxed side basins carved out when glaciers fell down into the main drainage below, in short it was perfect sheep country.

As the rams slowly moved lower down the hill, rain squalls continued, we squirmed and wiggled around behind our blind of trees for warmth. "Let's move further up the creek" Geoff said. There was a roll of glacial till on the side of the creek that provided us an avenue to steal further up the valley hidden from sight. As the rain fell we moved closer and closer. After another hour or so I poked my head over the slope and found the rams feeding on the grassy mound.

"Perfect" I whispered to Geoff, "they are all up on that patch of willows." As I watched, a couple of the smaller rams were beating the willows up with their horns, one or two stood on their back legs to get at the sticky buds at the top of the bushes. One or two stood sentry and a couple others were nipping at the grass on the ground. Across the valley the mountain goat was up feeding. Another squall came over the ridgeline above, a light breeze pushed the misty rain over us and down the valley, the rams straightened out in a line as they fed along.

Our partner Bill was up another valley, one we call *"goat valley"*, he had seen a couple distant rams high on a ridgeline there the day before and was up there trying to find them.

Geoff slid his pack ahead and placed his rifle over it. I set up my spotting scope and ranged the rams with my binoculars.

"245 yards," I said.

Geoff got himself ready while I starred through the grey mist to sort the rams out. The one Geoff liked was behind a few other rams in the feeding mix. In the gloomy day they slowly moved to and fro.

"He's the second to last one," I whispered to Geoff

"Got it," he said

I watched in my spotting scope as the group of rams moved ahead and stopped to feed.

Geoff's shot echoed back and forth through the murky air, the herd of rams spooked but stopped a few yards off to look back. His shot was perfect, the ram was down. We quickly gathered our gear up and made our way over to the ram. The other sheep slowly made their way up and over the top. I often marvel at how effortless mountain animals make a climb look so easy, in no time they were silhouetted on the skyline high above us. As more rain fell we set about taking pictures, skinning and deboning his magnificent ram.

The hike off the mountain that night was long and wet but with loaded packs it was done with smiles. We were soaked to the bone coming into camp that night but Bill had a nice warm fire going under the tarp, it didn't take long to get warm and start retelling the day's adventures.

Over the next few days the weather turned nice, the three of us hiked up and down a series of valleys looking for sheep. It was a magnificent time, we saw mountain goats, grizzly bears, wolves and lots of ewes and kids but no rams. One nice, brisk morning, I was up just as the sun peaked over the eastern mountain tops, I poured a coffee and went to sit in the sun and glass. Really I was just enjoying the sun however in a matter of seconds I found four rams high in the cliffs.

"Boys" I yelled, "get up."

I set my spotting scope while Geoff and Bill worried around camp getting dressed. Before long they both sat down and had their binoculars on the rams as well. Two of them were definitely short of the full curl requirement, one looked very close but the last one was legal for sure.

"I don't know Dawson," Bill said "that middle one looks good."

"Ya, that's my thoughts as well." I replied.

It was a beautiful fall morning that promised a perfect day to go after the rams. They were up a side basin fairly close so we decide to head out with empty packs and just one spotting scope. Our route would take us up a steep canyon and then up a rock slope to a grassy ridge. From there we should be able to locate the rams and make a plan from there. Bill was going to follow a creek further up the main valley and let the rams dictate where he would go.

A few hours later Geoff and I slipped over the ridge and sat behind some large rocks to glass.

"There they are" Geoff said pointing higher in the cliffs.

All four rams were bedded on a grassy shelf across the valley, there was no way for us to get at them from our vantage point. We were going to have to backtrack off the ridge we were on, go down the canyon, circle the base of the mountain then climb up the backside of the cliffs they were on.

"Dam" I said, "I wonder if they will stay there for the day?"

Geoff figured they would and thought the best plan was for me to take off and he would stay where he was to keep track of the sheep.

That's what I did, going down is always faster and before long I was down off the ridge, across the basin and making my way up the slope. It was a stellar day; a soft breeze blew gently in my face, the sun was warm but not hot, the steep slide I was going up was dry and easy to gain traction, I was in heaven. Everything was falling into place as I got to the crest and readied myself to look down towards where the rams should be. I could see Geoff across the way on the grassy ridge, he looked to be sitting behind his scope looking my way. I slipped my backpack off and slowly peaked over the cliff tops. The rams where bedded on the grassy shelf about a hundred yards below me. Quietly and slowly I pulled back and looked around, it was a perfect scenario, I loaded my rifle and slithered up to the edge.

As I set up one of the smaller rams got up for a stretch, the others started moving about. The big heavy ram stayed where he was, he swung his head to see what his cohorts were doing.

I sucked in my breath, *"what a beautiful ram"* I thought to myself.

Just as I was going to shoot he rocked forward on his knees and stood up, stretched and looked around. It's the sixth sense things wild animals have, almost like they know something is amiss in their world or that something is looking at them. The ram stretched forward and hunched his back, as soon as he stepped sideways he was broadside and I shot. The 140 grain nosler ballistic tip from my 7mm STW hit its mark and the ram fell off the bench. I watched him slide down the scree slope thinking *"don't roll, don't roll"* as he went. My worry was for naught, the ram slide without rolling and came to stop in a jumble of bigger rocks.

"Yehaa" I thought and quickly gathered my backpack up. I tried to go over the cliff right there but it proved unpassable. As the other three rams climbed

the jagged peaks with incredible speed and ease I made my way down the way I came up. An hour or so later I was just coming to the bottom of the chute when both Bill and Geoff showed up. "Whoo Hoo" Geoff said, "I was watching it when you shot" he said, he rolled right over. "What a dandy ram" I said.

Sheep hunting gets in your blood, it's a passion that's hard to truly explain, but, it's not for everyone and that is okay. Right there and then Geoff, Bill and I were three of the happiest sheep hunters around.

By the Skin of my Teeth

A Grizzly Bear Adventure

May your trails be crooked, winding, lonesome, dangerous, leading to the most amazing views and may your mountains rise into and above the clouds."

I suspect the late novelist Edward Abbey was referring to life's overall journey however his hypothesis from his book, Desert Solitaire, pretty much summed up our late September hunting trip. Over the previous two weeks we had experienced all the adjectives; and, as if we were following a script, there was a grizzly bear, loud, angry and wounded, running full tilt down the hillside towards me. I was down on one knee and had my rifle, with one shell loaded, cocked and the jittery crosshairs of my scope were trying to find a home. The bear hadn't known I was there until it came through the last batch of berry bushes about thirty yards above where I was kneeling. At the instant I moved its eyes and ears locked on me and the bruin altered its downhill race toward safety to where I stood. My partner Thane was on a little rise about four hundred yards away watching and videotaping the event.

We were indeed on a mountain; the trail had been crooked, winding, at times lonesome, and, at the moment, things were ramping exponentially up the dangerous scale. Lastly, I guess it could be said that Thane had a view that could be described as amazing……

Thane Davies, Bill Cash, Tom Johnson, and I were camped on a high plateau in northern British Columbia in late September. We were hunting caribou, moose and grizzly bear. The previous week had been rent full of adventures for all of us; the weather was perfect, cold in the morning with cool afternoons and evening, comradery was as good as it could get however our hanging pole was still empty. The colours, smells and sounds of fall in northern BC are exquisite, it's almost like nature is offering a cornucopia of visual, audible and olfactory treats before the long dreary days of winter set in. One can sit on a hillside glassing a valley and see all the vibrant fall colours, hear nature's characters prepare for the winter and smell the redolence in the air, it is one of the best times to be afield. Bill and I had hunted the area many times, it was Thane and Toms first time. Our routine was similar throughout hunting camps everywhere, we would get up early, have a cup of two of coffee, perhaps a porridge then load packs for the day and head out. Bill and Tom were focusing on swamp country looking for moose while Thane and I headed up to the high plateaus to look for caribou and hopefully a grizzly bear.

Thane and I were sitting in a high mountain pass that we had found earlier in the week. We had sat in the same rock pile the morning before and had seen a herd of caribou in the distance, there were a couple of bulls in the group but neither were legal. It was that and the sheer size of the area and the tremendous glassing that brought us back. We whispered back and forth in the predawn darkness anticipating seeing caribou, hopefully a legal bull would be in the mix this time. Dawn crept slowly towards us as we waited for better light. It was a crisp and still morning but the red tinge to the blossoming sky lent suggestions to what the day would be. As daylight provided we glassed further up the set of draws.

"Bear" I whispered to Thane

"Where?" he asked

"Third basin, right at the top," I told him.

Through my geo-vids I watched as the big bruin fed on berries high in the basin, it was classic spot for a grizzly bear to be. The bear was on a steep hillside covered with berry bushes, a few evergreen trees were scattered lower than the bushes and the bottom of the basin had a small creek running through it. High above treeline, quietly feeding in a patch of berry bushes with a noisy creek that should also draw the morning air down, the bear was in a perfect place for a stalk. Even at the far distance it was easy to tell this was a big bear, it was dark and blocky.

"How far is that?" Thane asked.

By cutting the distance in pieces and using the range finder in my binoculars I estimated the distance.

"About three and a half klick's" I said to Thane. "Probably be there by noon if we take off now."

We methodically unloaded our backpacks, took what was needed leaving the rest and stepped off. Our trail would take us down off the pass we were on, across and down a flat drainage then up through a spruce stand into the basin. Thane and I have backpacked together lots so quietly and with intent we set off, a few hours later we pushed over the last steep section into the basin.

"Let's stay up on the left side and just sneak through the trees" he said.

We couldn't see the bear, or where the bear was, but there was a rock outcropping at the top of the basin that we used for visual reference. Staying on the left side, part way up and sneaking from tree to tree we quickly closed the distance. We would stop by most trees, glass and whisper, the excitement was palpable. Almost to the top we stepped around a small tree and saw the bear down at the creek about two hundred yards away, it had moved down from where we thought it should be but was working back up. Most likely it had just came down for a drink.

"There it is" Thane whispered.

"Ya I see it" I replied.

The bear was moving back up hill towards the berries.

"Let's sneak in closer" I said as the bear moved away.

As we went down the hill Thane said he was going to find a spot to sit on and set up his video camera. I told him I wanted to get at least across the creek and hopefully sneak within a hundred yards. He stopped and I kept going.

As expected the rushing water took the breeze and any sound with it, using the sparse trees as cover I snuck closer and closer. The bear had reached the berry patch and was sitting on its rump raking berries. I edged closer. A couple times I stopped to look back to find Thane but couldn't see him.

"One hundred and fifty yards" I silently said as I ranged the bear from behind a small tree.

"*A few more trees closer*" I thought.

The trees were thinning out and if my memory was correct the next patch of three little trees would be the last before open alpine. I snuck closer.

The air was still on the bear side of the creek, a few late season dicky birds noisily fed and cavorted in the high basin. Arriving at the last group of trees I slid my backpack off and slowly, binoculars raised, peered around the tree. The bear was up to my left at about ten o'clock at about one hundred and five yards away. It was sitting on its rump feeding, quartered towards me. I slowly eased back behind the tree. Looking around I saw that to my right about ten yards away there was a hump in the ground. If I could crawl to it and lay my backpack over it I would have a perfect shooting position. Just before crawling over, as quietly as possible, I loaded a shell in my rifle. I had two more shells in my pocket and three more in my backpack. I was confident with one though. Thane and I had been shooting all summer and my Weatherby .338-.378 shooting a 225 grain Accubond was pushing tacks at one hundred yards.

I took one more chance to look back hoping to find where Thane was but couldn't find him. I got down on my belly and crawled, pulling my backpack along while keeping an eye on the grizzly. Any worries I might have had of being spotted were without merit as the bear was engaged in eating berries. I slowly pushed my back pack over the hump, laid my rifle over it and situated myself behind it. The bear was huge in my scope. It was still quartered towards me though so I watched and waited. I secretly hoped Thane was getting all this on tape, what an awesome adventure. A couple times the bear got up off its rump to shift to a better berry spot and both times I waited for it to take a step uphill to open up its downhill side, but both times it swayed left but didn't step left. Finally after about twenty minutes it stood up, I centered my crosshair. I watched through my scope as the big bear raised its head, looked over the high country, and drew in lungs full of air through its nose testing the airwaves.

The bear swayed uphill.

I found the spot with my crosshairs.

"Boom."

In the high basin the shot echoed back and forth, I watched in my scope as the bullet hit and the bear rolled sideways out of view. I stood up and looked back to find Thane but still couldn't see him. I was just ejecting the spent case when I heard.

"wshhhhh"
"whsss"
"crack"
"humphraaa"
"whsss shwshhhh"
"crack"

Turning to look up the hill I saw bushes moving and heard "hummmpha-humpha-hummmpha-rrrrrr."

Just as the bear came out of the bush I saw it. As soon as I moved its eyes locked on me; I remember clearly seeing its ears lay back, eyes narrow and its bearing, straight downhill, changed to four o'clock to where I stood. It was thirty yards away, running full tilt, loud, mad, wounded and probably just as surprised as I was. I placed a bullet in my rifle and closed the bolt,; went down on one knee, settled, well tried too anyways, the crosshairs on the running bruin and somewhere in my mind a thought said *"nope, you have one shot and one shot only, this isn't the time to use it"* so I stood up and thrust my rifle out like a spear. The plan, I have no idea where it came from or how I came up with it, was simply to wait until the bear was right on me, jam my barrel in its mouth and pull the trigger.

It was surreal, the sight, the sounds and the size of that bear at such a close distance (and closing so fast) was inconceivable. I didn't have to time to consider the predicament I was in, didn't have time to orchestrate a plan, or in fact, I didn't have time to be scared, I simply reacted.

Just as the beast got to me I skipped right, uphill, and jammed the end of my barrel in the bear's side pulling the trigger as I did. Its mouth was open wide open, teeth bared, ears laid straight back, the sight, smell and noise was incredible, its eyes followed my leap uphill but luckily for me its body didn't.

As the bullet tore through it the bear rolled over my backpack and piled up about ten yards below me, I was back peddling and trying to use fumbling fingers to find the last bullet in my pocket and load it, all without taking my sight off the bear. It gathered its wits and came up on its hind legs, its beady eyes bored into mine and roaring loudly with murderous intent.

So scary.

The big bruin lunged towards me as I let my last bullet fly. It rolled backwards but got right back up again coming towards me. I was out of bullets so turned uphill to run and heard "boom". Looking back to the bear I saw it roll over again as Thanes shot hit it. It was badly wounded but not dead, the sound amplified, the aggression shook the air. I looked uphill to run then looked back at my backpack which was halfway back to the bear. It was rolling around, roaring and tearing up the ground, the air shook with intensity. I am not sure where the nerve came from but I ran towards the bear, grabbed my backpack then ran up hill about fifty yards away all the time tearing at zippers to get my spare bullets out. I found them and got one loaded as the melee of noise and perspicuous energy settled down.

Loading another shell in my rifle I started to go down to where the bear was but chickened out and took a big wide circle to get to the other side of the creek and meet up with Thane.

"Holy crap" he said, when he walked up "I was up there wondering what you were thinking when I saw you stand up." "Oh man" he said "I would have crapped my pants."

"Well" I replied "I'm not sure I didn't."

We snuck over to a clearing where we could see the bear from and watched it for a while however it was for not. It was dead.

Thane and I sat down and replayed the whole event over and I told him why I didn't shoot from the kneeling position. We watched the video over and over while we marveled at the beautiful bear. I was an ordeal but a good friend of mine, and fellow bear hunter, Dave Phillips, always says "grizzly bear hunting is not for the weak of heart." His words could not be truer.

The trip out with heavy loaded packs was uneventful. Later that night as darkness took possession of yet another glorious day we went over the story time and time again with Tom and Bill around a crackling fire. Laying in my sleeping bag that night I replayed the event with the bear, I wondered where the decision to save my last shot came from and I marveled at the experience. I have hunted grizzly bears many times and respect the animal's size, strength and ferocity immensely. I wouldn't want to go through that experience again but after the fact felt fortunate to have gone through such, it was by the skin of my teeth that I came through unscathed.

Over the next few days Thane and I started seeing more caribou up on the plateaus, mostly scattered little herds moving through. We had stalked into range of a few however the herd bull wasn't what Thane had come to find. About noon of a bright blue blazer of a day I said, "let's head back to that pass and glass from there". We made our way towards the pass, glassing as we went. An hour or two later we were again in position and glassing the basins for caribou. Before long we had spotted a few lone or small groups of caribou, a couple required a better look through a spotting scope but were cows, calf's and smaller bulls. As the sun started its fall from the sky I glassed across the valley and saw movement up on a ridge. As I watched, a herd of nine cows and a magnificent bull came over the ridge line.

"There's your bull" I said to Thane, pointing to where the herd was.

Putting the spotting scope on it only proved the point, it was a royal bull, white maned with a high and wide rack. Thane looked over and smiled "let's go" he said.

As we made our way down off the side we were glassing from the herd started moving further up the box basin. I was worried they might keep going right over the top but once into the steep end they stopped on an island of grass in the scree. We just kept hiking and soon were within nine hundred yards. Closer and closer we moved up the mountain, finally dropping off the back side of the ridge to hike out of sight. About three hours later we crawled over the crest and settled in behind a little divot on the hill side. We were six hundred yards away from the bull. It was wide open and we couldn't get any closer, the sun was falling right behind the ridge above herd, there was a storm of dust and flora particles dancing in the sunlight.

Thane pulled his backpack up and set his rifle over it, he had a perfect prone shot.

All the caribou were up feeding, the bull was second from bottom. Thane looked at me and asked the distance, I said "six hundred yards dead on." He referenced his cheat sheet and fiddled with the turrets on his Vortex scope. The bull changed positions to feed up hill. He got in behind his rifle again. He and I had been shooting at gongs well over six hundred yards all summer long so we were both pretty confident that he could hit what he wanted at that range.

"Get ready" I said as we watched the bull move positions to be completely broadside. Once the bull stooped to feed, broadside, I ranged again to make sure distance hadn't changed and got behind my spotting scope again.

"Anytime" I said.

I watched the whirl of the bullet cross through my scopes vison and hit the bull right behind the shoulder.

"Perfect shot" I whispered to Thane, he looked over and smiled. The herd jostled around on the steep slope wondering what the noise was, the bull swayed but didn't go down. Through my spotting scope I could see the crimson blotch grow on his side. We watched as he staggered but didn't go down. "Maybe hit him again" I said. "Okay" he said. Again I watched as his shot found home and this time the big bull fell down and rolled over. As the bull rolled down the cows took off across the slope.

It was almost dark by the time we got to the bull, just time to snap off some pictures, field dress it and break it down. We left the quarters and hiked off the mountain taking the horns and enough meat for diner with us. It was a steep, dark journey out of the basin and off the mountain that night. Around the campfire that night Thane told Tom and Bill about the adventure, it was a great hunt.

Thane and I went back to retrieve the caribou the next morning while Bill and Tom headed off to a swamp not far from camp. A few loaded packs and half a day later we had the caribou meat off the mountain and were back at camp hanging the quarters.

Just as the sun began its decent Bill and Tom came into camp with the news that they had killed a bull moose in the early afternoon. We quickly finished hanging the caribou and went with them to look after their bull.

It was another fun night sitting around the campfire reliving the events of the day. Bill and Tom's bull was the final chapter to an exceptional hunt. We were taking home fantastic memories of our trip, incredible stories of adventure and meat for our tables.

Is Today the Day?

A Wolf Hunting Adventure

"I think today's the day" I said to my partner Bill Cash.

"You think so" he replied, poking his head up out of the sleeping bag.

Small clouds marked our breath, it was cold, real cold. The thermometer stood steady on the minus thirty two mark, a fresh wind blew from the north. Small whirlwinds of snow danced here and there on the lakes frozen surface. Across the valley the first golden rays of sunlight slowly crept over the snow-covered peaks, high in the air ice crystals pranced and glittered in the burgeoning sunlight. It was indeed going to be a good day. Bill got up and stoked the small wood stove. Slowly the cabin shook off its icy gloom and warmth spread throughout as the fire roared to life.

Bill and I were in a little log cabin on the shores of Trimble Lake in the Sikanni Chief water shed. The cabin we were in as well as the other ones we had bunked in the past five days or so belonged to Mike and Dixie Hammett of Sikanni River Outfitting. We had met Mike and Dixie as well as their son Zach, daughter Jenny and wrangler/guide Clint the year before when we explored the drainage on skidoo's hunting wolves. That trip didn't produce any wolves but we did see a few, and saw lots of sign. Though we came out empty handed that time it was the first destination on the wolf season at hand.

A bitter wind pushing minus thirty temperatures greeted us as we pulled to a stop in front of the main cabin at the Sikanni River Ranch. The ranch sits quaintly in the crux of a corner on the north side of the Sikanni Chief River in north eastern British Columbia, approx. 170 miles up the Alaska Highway. Clint poked his head out and greeted us. Mike, Dixie and the rest of the Hammett's were away in sunny Texas while he stayed behind to tend to the stock. Before entering the cabin I happened to look aver at the skidoo and sled behind it parked by the entrance, in the sled lay a beautiful gray wolf. Clint explained over steaming mugs of coffee.

"I was sitting right here when the pack crossed that clearing" he said, gesturing to a field a thousand yards or so away. "I just grabbed my gun and took off on foot to head them off. The main pack got ahead of me but this one and another straggled behind" he continued, "I shot this one and missed the other."

Bill and I looked at each other, this is going to be great we said in unison. Clint agreed "you guy's should do well" he said, "a large pack headed up stream just a few days ago" while he couldn't say for sure how big of a pack he did say that it was a big one. We reminisced and passed hunting stories back and forth before retiring to one of the smaller cabins for the night, Bill and I talked about the upcoming week well into the icy night. The wolf in Clint's sleigh as well as his words about the pack up trail had us excited.

    The Sikanni ranch has line cabins placed along the trail about every twenty five miles or so (approximately a day's horse ride) along the Sikanni river. Our plan was to skidoo up to the first one big mountain, leave most of our gear except for our food there, and head further up the trail to the blue bell camp. There we would assess the sign or lack of and decide where to concentrate our efforts. Wolf hunting, at least how we do it, is very similar to elk hunting. We would locate them with loud calls then move in close and seductively call the animals into range. Bill and I are avid wolf hunters, we spend at least a few weeks every winter hunting them. Wolves mate in February/March and that is when they are most susceptible to calling, very similar to the peak time's moose or elk respond to calling. January is a little early but we were both itching to go so we went anyhow.

    We traveled along that first day, stopping at valley mouths, high spots and river mouths, calling as loud as we could. As our long wolf howls drifted into the reaches of these abysses we would strain to hear a reply. There was abundant sign of the pack along the trail on the creeks and rivers but we failed to get a single answer that first day. Pushing trail further we got to the blue bell camp just as dusk settled over the pristine snow covered landscape. The moon cycle was at full peak so darkness was really only a shadow filled daylight, we both thought that the wolves would use this advantage and be at peak activity.

The next day we again set off, howling our mournful calls here and there. Other than seeing some moose, caribou and the ever-present buffalo we didn't have any luck locating wolves that day either. Even on a bad day, a day spent winter wolf hunting is still a great day. With snow as a canvass, animal markings tell a distinctively different tale than those usually seen in the dirt and mud. Frozen waterways become natural highways. We traveled up and down creeks, rivers and valleys that would be impassable in the usual fall season. Sign in the snow, whether it is ungulate tracks, small predators like wolverine, lynx or just birds cavorting offer a totally new insight into the resident animal's daily business routine.

By dusk that night, another night that offered near daytime light, we had settled into the small big mountain cabin. Bill thought that we should venture over into the Trimble lake country, from there he mused, we could head up to Redfern lake, the Besa river drainage and even down the Nevis creek valley. I agreed with his thoughts, nobody had been into the country yet, it was just early January. We might have to push through some deep snow here and there but it should be fun. We had hunted moose on horseback in the Nevis creek side the previous fall and thought it would be neat to see the area wearing its winter robe.

The next morning we lashed a food basket into the back of Bills machine and turned deeper into the winter wonderland. At the north-west end of Trimble lake we continued up to the valley's end and met with the trail that accesses Redfern lake via the Nevis creek waterway. We decided to head further west towards Redfern and had only traveled a few miles when the trail we traveled on was covered in wolf tracks.

"Aha" Bill said standing beside his machine as I caught up to him. He pointed towards the tracks,
"this is starting to look more like it."

We pushed on with renewed enthusiasm. But the wolves didn't cooperate with us, they simply didn't or wouldn't answer our calls, or maybe we were in fact just too early to really get them going with our calls. We went all the way to the far end of Redfern lake and failed to rouse even a peep from the timber wolves we knew to be in the vicinity.

Bill suggested we head back and travel down the Nevis creek drainage, "there should be a ton of moose in that drainage" he said, "so maybe the wolves will be over there."

And they were, but unfortunately just not when we were, their tracks completely covered the trail. At a few spots it could be seen were they actually had laid around on the trail, "soaking up that warm sun" I said, "probably so" Bill replied. However, the moose were there, there in numbers we didn't quite expect, from one spot I counted close to fifty cow. Another hundred yard down the trail and those fifty cows turned into probably seventy, there was moose and caribou everywhere. Bill said that nobody would believe the sight, they were spread across the valley bottom like domestic cattle on a farm. "Now we know where the wolves hang out Dawson" he said, "you betcha, and now we know why" I replied.

The thirty mile trip back to Trimble lake that night was short and sweet, anticipating the next day's events we lost ourselves in the moon lit ride through some of BC most awe inspiring mountain views. Again we sat around the plywood table planning the next day assault. Trimble lake boomed and cracked from the dropping temperature, I lay wide-awake in my sleeping bag anticipating the day ahead.

"So ya think this is it eh, today's the day?" Bill asked while stoking the wood stove.

"Well it's either that or drown ourselves in pessimism and screw up the day right now" I replied chuckling into my coffee. "Actually" I continued, "I think that with all the moose and caribou over that side the wolves must be hanging close, what we need to do is get right on over there and start hollering, and as usual a small serving of luck would help."

Without a seconds thought Bill went out and flashed up the machines.

Another sunny day blossomed to life as we headed down the lake, a familiar group of five cow moose greeted our presence, as they had every other day, in the frozen swamp at the lake end. We quickly retraced our trail over into the Nevis valley. We called without luck the full length down the drainage. Stopping at the end, we shut off the machines again.

"We'll what now?" Bill asked.

"Let's head back up, but follow the creek up" I said.

At one high point along the creek we stopped and called, the echoes floated lazily through the frigid air. Bouncing off a mountain they would come back to us, float off to another drainage up over there and come swirling back again. It was a natural amphitheater, the acoustical quadrate was near perfect. I couldn't imagine that there was a wolf within twenty miles that wouldn't hear our howling rhapsody. But again nothing responded. Coming around a corner on the creek I saw a covey of ptarmigan resting on the ice. I stopped and shut down my skidoo, "what's up" Bill hollered.

"I just want to take some pictures of these birds" I called back.

Bill shut his machine off and I clicked away at the snow-white birds. Without the black markings on their wings these birds may have come up with nature's best camo. A few waddled slowly away and simply disappeared against the white background. Bill let out a wolf howl.

That's when the magic happened, did luck entered the equation? Or was it our persistence? Whatever it was, we didn't care, as we stood on that frozen creek looking at the ptarmigan a mystical sound floated from afar to us. We had traveled close to three hundred miles in the past week just to hear the sweet chorus of a wolf pack in full song, and they were in fine form. Howls from the adults intertwined with barks and yelps from the younger ones. Nothing could have been sweeter to our strained ears.

"There" I said to Bill pointing to a draw just across the valley.

"Unreal" he said, "we passed within spitting distance of that stand of trees an hour ago."

"Let's jump over to that small hill and get on top" I said.

Hesitantly I started my machine, I felt like the sound from it would be heard on Mars. I was sure, positive, that the noise would scare the pack away.

But we reached the small rise and the pack was still singing, I called back and received a chorus in return.

AHHHHOOOOOOOAAAAAahhhh, I called

Ahhoooaaaaaaaaahooo, they responded.

"Grab your gun" I called to Bill, "their coming out."

We reached the top and settled down to wait as the wolves continued the noisy vocalization. I called they answered, we strained our eyes through shaky optics looking for them, I was sure that they were inches behind the forests cover straining their eyes trying to figure us out.

The calling back and forth went on for about fifteen minutes, suddenly without reason they quit, so we did. The valley went still, the air vibrated with anticipation of the next scene in the drama that was unfolding within its frozen grasp. The pack must have used the time at hand to move position. Then the calling erupted again shattering the stillness that only moments before had stilled even the wind, they had moved closer. I was just completing a loud mournful, I need a wolf girlfriend yodel when Bill whispered "there they are."

"Where?" I asked.

"Over there" he pointed but I couldn't find them. He had seen the pack run between a stand of trees.

"Give em another call" he said. I called, nothing, I called again, and another nothing.

We were starting to wonder if they hadn't taken off when I said to Bill, "I'm going to call, you jump into a call a second or so after I start", "Ok ready" he said.

Ahhhhhhhhhhhhhhooooooooooaaa I yelled

Ahhhhaaaaaaaoooooo Bill called

Bill started a second after I did and we finished almost together. That was enough, I guess that was the last straw as barely had we come to finish as a string of timber wolves came single file out of the pine forest heading straight for us.

"Here they come, get ready" Bill said in an excited voice.

We settled behind our rifles and waited. They closed the distance, we waited. They trailed one after another through a small cluster of trees. As if choreographed and rehearsed by some earlier experience, or by instinct, they fanned out without missing a step presenting a strong front as they passed through the last cover between them and us.

Five large and apparently all male wolves, one pitch black the others blond and gray came to rest on their haunches about five hundred yards away.

"Look at that" Bill said. "Wow" I replied, "I'm going to shoot the black one" I said, "you ready?"

"Not yet" Bill said, "how many are there."

My scope was cranked up to nine and I slowly counted across, "there's five" I said, "the black one is center, two on either side."

"Ya I see the left ones but not the other two, oh wait, yup now I see them" he whispered.

"Okay" I said. "I'm going to take the black, no wait, the black ones up, he's crossing behind that huge gray one on the right, oh crap, I think he's leaving, nope wait he's stopped, okay ready" I asked Bill.

The wolves were starting to mill about.

"No wait a second, I lost the one I want" Bill said.

All the wolves sat down on their haunches showing a strong unified front as we panned back and forth with our scopes. The black wolf had got up and paced back and forth behind a couple others a mili second before I was going to shoot. The others had looked at it and then turned back to us as I blasted off another call. It was almost as if the black one knew he stood out better than the others as he would move then stop, stutter a second then move again the other direction, stop behind a bush then move again. The four light colored ones were almost invisible in the flat light. If we hadn't watch them come down from the forest I question if we'd even know they were there. The black wolf looked back as if to go back to the waiting pack. It was a truly remarkable scene, one that must be experienced to fully grasp the opportunity we had that afternoon. As I sit here I'm left with the thought that in the end our duel call must have convinced the pack that there was two lone males looking for mates, perhaps the five males were sent out to either fight for the packs integrity or chase the intruders out

"They are getting ready to take off" I said to Bill, "you ready, "which one you shooting at?" he asked.

"I'll take that big one on the far right" I said, "you take another, ready, on three" he said. "One.. two... three."

My .25-06 thumped against my shoulder, through the scope I saw the wolf flip over. Bill was a second or two late, at my shot the wolves were up and running away. Bill tried at a running wolf but they were too far and moving too fast to realistically make a shot. The pack quickly faded back into the safety of the forest.

"I got one Bill" I cried, "man that was awesome" he said. Just then the pack erupted in a frenzied crescendo calling their missing member, that particular moment will stand out as one of the most remarkable moments in my life. I swear the air moved, it was unbelievable. Bill and I looked at each other and moved ahead hoping to get another crack at the pack. We called, they roared, we called again, they roared louder and louder. As we moved and called they increased their volume and activity.

As fast as it all happened is as fast as it quit. The leaders must finally had enough and led the pack out the back door and high tailed it. Bill and I made our way over to the fallen wolf. It was a huge blond and gray male. We both wondered if it was the alpha male, *"surely there wasn't any bigger ones"* I thought. Both Bill and I had harvested wolves before but the particulars that this hunt provided were exceptional. When we got home a few days later he weighed the big male wolf. Exactly one hundred and twenty pounds, an awesome trophy, an awesome experience. To have that interaction with them and to call them out was perhaps one of the best hunting moments I've ever had.

A full silver moon hung like a globe in the serene sky as we turned back towards Trimble lake. There wasn't a moose or caribou to be seen, not all that surprising after the vocal aggression that engulfed the valley the preceding two hours. I told Bill that if they were smart they would be hiding or gone to friendlier feeding grounds.

Tales from the Trail Photo Album #1

'The best thing about hunting and fishing,' the Old Man said, 'is that you don't have to actually do it to enjoy it. You can go to bed every night thinking about how much fun you had twenty years ago, and it all comes back clear as moonlight.'"

Robert Ruark

Bill and his horse Mr

Dawson with the grizzly bear
"By the Skin of my Teeth"

ADVENTURE STORIES

Geoff and his ram,
"Hard Bargain Valley"

Bill and Ellie, we always had to cover her
eyes to pack game on her

Dawson on a ridge spotting sheep

Wolf Hunting

ADVENTURE STORIES

Bill standing by a grizzly bear marking tree

TALES FROM THE TRAIL

Dawson with wolf
"Is today the Day"

Thane hiking to our glassing spot
"By the Skin of my Teeth"

ADVENTURE STORIES

Bill on Mr leading our packhorse Cisco

Geoff and Dawson, on ridgeline sheep hunting

TALES FROM THE TRAIL

Dawson with mountain goat
"September Sounds"

Bill and his grizzly bear
"Redemption at Windy Ridge"

ADVENTURE STORIES

Bill and Dawson with Dawson's ram
"Hard Bargain Valley"

Dawson with his horse Star loaded with his ram,
"Redemption at Windy Ridge"

TALES FROM THE TRAIL

"Hard Bargain Valley"

Bill with wolf
"Is today the Day"

ADVENTURE STORIES

Dawson, those are grizzly bear tracks crossing
the snow, "Hard Bargain Valley"

TALES FROM THE TRAIL

Bill in one of the cabins
"Is today the Day"

Dawson across from
Black Sand Mountain

ADVENTURE STORIES

Sun setting on "Hard Bargain Valley"

Bill with his horse Mr and our
packhorses, Cisco and Eagle

TALES FROM THE TRAIL

Bill along a little lake
"Redemption at Windy Ridge"

Dawson and his cougar
"Way Too Close"

ADVENTURE STORIES

Hard Bargain Valley

Dawson and his ram
"Redemption at Windy Ridge"

Just when you think you found a spot where no one has ever been you find a can in a tree!

Bill and his horse Mr with his grizzly bear, "Redemption at Windy Ridge"

ADVENTURE STORIES

Dawson on his horse Star with packhorse Ellie, Claw Mtn.

Thane with his bull caribou
"By the Skin of my Teeth"

TALES FROM THE TRAIL

Bill, we called a black wolf right
down the creek to us

Dawson looking for rams
"Hard Bargain Valley"

ADVENTURE STORIES

Thane and Dawson glassing from rock pile, "By the Skin of my Teeth"

Dawson with the grizzly bear "September Sounds"

Bill with the black wolf we
called down the creek

Dawson with his horses Ellie and Cisco.

Pronghorns in the Desert

An Antelope Hunting Adventure

Webster's Ninth Collegiate dictionary describes hunting as "a continuous attempt by an automatically controlled system to find a desired equilibrium condition." For myself that means the year in year out search for the holy grail of hunting; the proverbial field of dreams, like the easter bunny, tooth fairy, and yes even santa claus, the holy grail of hunting is of a fleeting uncertain and I guess even questionable identity. But the search for it is what powers most of us through the desperate days of closed seasons. The neat thing about my personal search is that the holy-grail keeps changing and the days of closed seasons are fewer and fewer. Last year it was mule deer, black bear, mountain goats and bighorn sheep. This year stone sheep, goats again (we all have our favourites) whitetail deer, moose and elk, next year all the above plus grizzlies, caribou and on it goes. Just when things have reached a calm the mailman brings news of distant hunting opportunities previously not thought of, ah...

And so it was that mid-September found my partner Bill Cash and I flying towards Albuquerque, New Mexico. The land there promised fields of heavy horned antelope. It could be said that hunters have a second chance of the naïve believing days of youth but that's for another time. For the moment at hand our quest was fresh, we were back on the continuous attempt trail and loving every minute of it. Now, the naysayers could point out that neither of us had ever hunted the black and tan bucks before, or even actually seen a prairie goat outside the tattered pages of a hunting magazine. But they couldn't say that New Mexico wasn't known to hold trophy busting bucks, at least that was a sure thing. They could say all the negative things they wanted and more but as we sat on the plane with a crib board between us on the center seat we didn't care. 15-2, 15-4, 567...

Adventures are like that, some are better than others but all better than none...

The good part about living/hunting in north america is that there is always another game animal to help one reach the desired level of equilibrium intoxication. It's the doing that gets the blood boiling not just the shooting...

Landing in New Mexico we were met by my friend John Harlan who lives in Albuquerque. John had called me the previous January with an invitation to join him on an antelope hunt. The state of New

Mexico has a permit/draw system for their management of game. While getting our non-resident hunting licenses a New Mexico game warden explained how the yearly permits were allotted. State game officials fly over a specific management unit head counting antelope, armed with this information they then come up with a harvest figure, half of the allotted permits are placed in a draw system while the other half are given to the rancher or ranchers who own the land. These landowner permits are managed by the landowner; John had acquired three of these permits for us from Donny Ansley owner and operator of the sixty square mile Red Canyon Ranch in central New Mexico.

Dust bellowed up as pulled into the ranches headquarters, across the yard strode a cowboy. As his boots hit the dusty red dirt another dust storm billowed up, the tall wrangler clad rancher sauntered across the small yard towards us, his gate, clothes and look all spelled the unmistaken toughness that comes from a working cowboy. With the unrelenting ninety-five degree sun at his back his image was one straight out of western movie. I was half expecting him to say draw, but we all know he didn't.

"How y'all doing?" he asked in a deep southern accent. "Hey John (it came out as Joun) who're these here boys?"

"Hey Donny, how's it going" John answered "these are the guys I told you about, Dawson Smith and Bill Cash, they came down here all the way from British Columbia."

"All the way from BC" he replied.

"Yes sir" John answered.

Donny's steely weathered eyes looked us over "that so boys" he said as we introduced each other and shook hands.

"I was up in BC last fall fishing, neat place y'all come from," he said.

Donny invited us inside his ranch house and showed us a couple of head mounts from the previous season. Bill and I exchanged a giggly crazed look *"its real man, we've found nirvana our thoughts spelled out."* Thankfully neither Donny nor John saw us or I bet we would have flew out of Albuquerque the same day we flew in.

After a few minutes in the ranch house Donny, John, Bill and I drove around to a high mesa, from there we talked and glassed distant antelope, or goats, as they call them down there. Donny pointed out landmarks and explained the layout of his huge ranch. I was elated, there were antelope seemingly everywhere. Bill and I were like kids in the proverbial candy store.

The actual season wouldn't start until the following Saturday so we had three days to scout the land. Heading back to town John explained the way to score antelope and we planned the next three days scouting. They were a blur, we would head out to the ranch arriving at dark, driving around we would spot distant antelope and set up our spotting scopes; by the end of the first day we had seen hundreds of antelope, the bucks were busy chasing does, apparently the rut was in full swing.

The next day we crawled closer to a few bucks that appeared to the bigger ones we were looking for,

"See the heavy bases and tall horns on that one" John whispered to us, "his ears are about six inches so he should be about sixteen inches plus."

In one deep arroyo (gully) we spotted what appeared to be a big mature buck, he was busy chasing a whole herd of does around, sneaking in for a closer look I could make out tall heavy horns, when the buck looked away his horns dropped to make the desired heart shape, both tips had ivory tips. My heart pounded.

"That's the one John" I said as we snuck out of the draw, "he's real good."

"Oh ya" John replied "that's a dandy buck."

The only thing that worried us was that Donny had twenty six other hunters who would be on the ranch for Saturday morning. We had spent three days scouting and had, what we thought, the three best bucks all figured out for the two-day season. John, Bill and I had found two awesome bucks way in the back corner of the ranch, both were big, high, heavy horned bucks. One was in a position were a hunter could make an approach on the backside of a brush covered mesa. The other was out on an open plain, the only way to get to it was to crawl a half-mile or so in the heavy chimisa or sagebrush. Both were do-able and both appeared to be trophy class animals.

By Friday night we were fairly certain what we planned for the following morning would work. Donny had invited us back to the ranch house for dinner so we headed back as the hot sun slipped over the horizon.

We were introduced to a handful of other hunters in for the weekend, Joe, Robert, Bobby, Arman, John...Doug another John, so many guys it's hard to remember all their names. Usually Bill and I hunt the real remote BC backcountry with a pack strung on our backs or with horses, neither of us had encountered a social evening such as that in a hunt camp before. It was a fun evening sitting around on the porch swapping hunting tales and half-truths with a great bunch of guys. Of course everyone asked where the other would hunt the next day, and of course everyone told the others as that they were going to hunt in a spot as far away from where they would actually hunt as they could get while staying on the ranch. We laughed and finally someone said we should just wait until Sunday to tell each other where we were going to or had hunted.

On the way back in Friday night we planned for the next morning. Our selected bucks were miles apart on the ranch, we decide that they would drop me off on the highway a few miles from where the buck I liked was hanging out. John and Bill would drive around to the far end of the ranch and each take off after their bucks. It was a great plan but I still worried that someone would drive into the area and scatter my buck. Donny had assured us that most of the hunters wouldn't be where we planned to hunt but I was still a little worried that either someone would find our spots or bust the bucks from their locations.

After three days of sweltering heat, staring through heat wave blurred optics at distant bucks it was coming down to the anticipated time. I don't think I slept more than five minutes all night. My alarm blasted on alert ears at four thirty am.

"Come on Bill" I said let's go. John was up, we were cranked.

I got the impression that John was worried about me heading across the couple miles to where I hoped my buck was still hanging out.

"Don't worry" I said, "I'll be okay."

After they dropped me off John asked Bill if he thought I would be okay. Bill told John that he wouldn't worry about me for at least a week, we have bushwhacked in real tough country before.

I quickly headed across the open desert heading for a distant mesa, the day before we had spotted a flashing red light about ten miles away. From where they dropped me off I could head straight for it and come out above the antelope. In the pitch black I hiked along. On the distant horizon a morning glow slowly crept over the landscape, other that walking into a couple prickly pear cactus and one startling encounter when a barbed wire fence stopped me, the hike across was a piece of cake. A hill ahead of me looked to be the backside of the one I wanted to be on. Crawling up to the top I tried to stay close to bunches of sage brush *"just a few yards further"* I coached myself.

Stopping behind a thick brush I scanned the open prairie that I could see. Nothing was in view. I crept a few inches closer and again stopped to scan as the basin revealed itself. All my attention was riveted on the emerging land below as I slowly, inch by inch, crawled ahead. I had just moved forward when a ch.. ch..chchchchhchc erupted like a siren from the bush beside me. I jumped, no, explode, off the ground, over the crest of the hill into the basin below. Stopping to gather my breath, nerve and to find my heart, I noticed that there wasn't any antelope in the bowl. Looking harder I saw that I had another half mile to go to the next mesa. Man, that snake or whatever scared me, my heart was pounding like a freight train on an up-hill track. By the time I reached the next hill I was back to normal. Not being any smarter from the encounter I dropped my daypack and started crawling up over the hill. Slowly I crept over the slope of land and peered through the brush into the bowl below.

A doe antelope shined in the morning sun, another, hidden by the morning's shadow, fed away from me. A familiar tickle raced through my body, a small shiver shook my body. Just fifty yards away on the other side of the bush I lay beside stood the grand buck. His deep black horns looked like primeval weapons attached to his head, the sun fell on half of him the other half was dappled by the slowly moving shadow. The early morning's sun progress chased the shadowy darkness from the basin. Another doe walked between the buck and me, another out on the slope stared right at me. I hoped my camo clothes hid my presence. She finally fed away, satisfied that nothing was amiss. The buck stood proud and regal a short arrows flight away. I had originally planned on hunting with my bow but opted for my rifle instead, I was kicking myself for that decision right then. The buck walked off towards a feeding doe, she obviously didn't like his advances as she took off like struck lightning with him in chase. Back and forth they ran, small clouds of dust trailed their path, him herding her like a champion quarter horse, her dodging and weaving. The other antelope continued their feeding. I took the opportunity and pulled my rifle ahead of me. The doe soon tired of the chase and settled down about one hundred yards away, the buck stood off to her left in dominant king of the prairie stance a short fifty yards away.

I often wonder why the seconds leading to the accumulation of a plan seem to roll by in slow motion, or is it just me. My Remington 700 launched the 150 grain nosler ballistic tip on its way. The once defiant dominant, grand beast dropped to the ground. I slowly eased up as the does milled about looking from one to another for direction. Once I stood up they headed for distant pastures. I quickly retrieved my daypack and went over to the buck. I like hunting alone but usually wish for someone to share that special feeling a hunter gets when they have executed a perfect stalk and made a humane kill.

Under the morning sun on a shadow filled mesa five thousand miles from my home I treasured the special moment and the special animal I had just taken. My desired equilibrium condition had been found, the continuous trail had an end to it after all, or perhaps, a temporary resting spot.

A few hours later, after I had taken a bunch of pictures and caped out my buck I saw John up on a distant hill in the truck. He quickly made his way down.

"That's was awesome" I said shaking his hand.

"Congratulations" John said, "What did he measure?"

"Sixteen and a half inches" I replied.

We talked about the stalk while loading the animal and my gear into the truck.

"What about you and Bill?" I asked "did you guys get to them bucks?"

"No" he said, "the buck I like has crossed over into the next ranch, Bill is now stalking that other one."

We slowly drove over to where Bill was supposed to be. Parking on a distant hill we set up the spotting scopes and looked over the land. Bill wasn't anywhere to be seen, I saw the buck John was after a long way off and defiantly on the neighboring ranch. Spotting another decent buck John and I headed off for a closer look. It turned out to smaller than we thought so we eased back and again looked for Bill.

A small herd of antelope with a real good herd buck was just over the fence line about a mile away. Unknown to us Bill was nestled in a thick stand of bush about fifty yards from them, he had snuck into them hoping they would cross over to our side of the fence line, but they were hung up on the wrong side and looked to be staying there. At about noon, with the temperature hovering over ninety degrees, he finally had enough and stood up. The goats sped away to safety as a half-cooked Bill made his way up to us. We decided to take my buck back to the camper, ice the meat and cape have some lunch and come back for an evening hunt. I would join them and help spot. We saw some descent bucks that night but nothing that Bill or John wanted. The desert sky slowly darkened as we made our way back to the ranch house for dinner. Many of the hunters on camp had taken bucks that day and everyone was talking and reliving stalks over a delicious meal Donny's wife had prepared.

Sunday morning – we were again in the distant part of the ranch where the two big bucks hung out. Just after light we spotted one still over on the other side of the fence but the other had crossed back to our side during the night. Bill quickly headed for an arroyo to start a stalk, John and I stayed where we were and watched Bill. The big buck was in a group with two does and six smaller bucks. As we watched Bill crawl towards them the bucks started chasing the does around. The big buck and a doe stayed put but the other bucks were chasing the other doe all around the mesa. They were back and forth slowly working Bills way, the doe with the big buck took up the chase and the big buck followed. They were all working back and forth closing the distance towards Bill. John mentioned that they might work their way right into where he lay in wait. The big buck herded his doe off to the left and came right past Bill, from our position high on the hill it looked like they were going to jump right on top of him. They kept coming and coming and went right by Bills position.

With amazing speed they came right up to where we hid. The buck hesitated on a slope about seventy five yards away. Johns .25-06 roared in the still prairie and the buck fell. The other antelope continued there frenzied chasing of the does as if nothing happened.

As quickly as they came, the other antelope sped away. We all met at Johns buck, it was an impressive animal. I pulled out my tape and it stretched to fifteen and a half inches with good heavy prongs. By the time we had it loaded in the pickup the prairie was again undisturbed. While John capped out his buck Bill asked me to drive into a far pasture and leave him there, before we got there he spotted a good buck over a small rise.

He jumped out and worked over towards it. Before I could even see it in the spotting scope his .270 barked and another buck lay on the dusty desert ground. Gathering my daypack I followed Bill's hollers and walked up to him and his buck. It was a dark faced buck we had seen the day before, his heavy horns stretched the tape to fifteen and three quarter inches.

Bill and I grinned like schoolboys as we boarded the plane to go home. In our baggage were two excellent trophies, far better than we had anticipated. We also were taking home some unforgettable memories of pronghorns in the desert and some new friends and memorable times with fellow hunters. Our search for optimum equilibrium levels in New Mexico had turned out allot better than either of us ever expected.

Out came the crib board and Bill dealt me three fives and a jack. As I grabbed the cards to cut the deck I felt the twinge of anticipation... a five... a sixty inch bull moose......a one ninety class whitetail. The season was young do I dare cut the cards and blow all my luck this early in year...

Ah, a three ...

clas-sic \klas-ik\ *adj*: noted because of special literary or historical association

Merriam-Webster.com

Hunting mule deer during the November rut season is akin to northerners hunting moose in September or bird hunters honking in flights of Canadian geese on Saskatchewan's wheat fields. A number of fine magazines, like Big Buck, focus their entire publishing futures solely on the legendary deer. Many early editions of the pioneering hunting magazine Outdoor Life had incredible apple tree like horned bucks gracing their front covers. I remember as a kid reading my Dad's dog-eared copies of the renowned magazine. Back then mule deer seemed to be the predominant big game species outdoor writers wrote about. Their stories of hunting big bucks on horseback high on mountains in strange and mystical places like Montana, Colorado or Idaho held me spell bound for many a nights as I read and re-read them with a flashlight under my blankets. Because of that or perhaps due to some great adventures I've had hunting high country bucks since then a fall mule deer hunt is tops on my annual to do list.

"Odocoileus hemionus" on Horseback

A Mule Deer Hunting Adventure

"Hey Bill" I asked, "do you believe in good luck things?"

We were driving west on highway 20 having left Williams Lake British Columbia a few hours earlier. He looked over at me, taking his eyes off the road for a minute.

"Like what?" he asked.

"You know, rabbits feet, lucky pennies, special trinkets, amulets, those kind of things?"

"Ah, ah...umm I don't know" Bill replied.

"Well I do," I said solemnly, "this trip is going to be great, we are going to find humongous bucks up on those plateaus. I can feel it."

"You always say that" he replied.

"Ya I guess you are right, but this time I've had a premonition" I said blatantly.

"Oh boy, not this again" he whispered into the steering wheel, not looking at me. I heard his muted lamentation just the same.

Over the next few hours I told him the tale that led me to believe that our adventure at hand would surely be ordained by friendly ghosts from the past. As we rolled along through the rolling sage and pine tree country, our motor home and horse trailer eating up mile after mile of the black ribbon road like a mechanical turtle, he listened as my tale of grandeur came to life.

"You see" I continued, "years ago my buddy Thane and I backpacked all over these mountains looking for big bucks. The first couple years were flops, we either didn't see anything good or we didn't see anything period. Bad weather beat us a few times, sheer bad luck a couple more, many a sound-minded men would've given up but we went time and time again. The last time, about a week before Thane and I left to go hunt the mountain you and I are heading for right now, my daughter Macenzie made me a neat little necklace. She called it my *"goob lup netlice"*. You gotta remember that she was only two at the time and the necklace was just a bead on a string. To her it was as good as any neck ornament ever made, to me it was priceless. At any rate, off we went, I was wearing my new necklace. The hike up the mountain that year seemed easy, our backpacks seemed lighter than usual, the sky was as clear as I had ever seen it. We made camp that first night and hit the sack. The next day and for the next six days after, we saw deer everywhere, the weather was perfect. Somewhere along the trail on the sixth day I lost the necklace Macenzie had made me. I'm not sure but I think it got snagged on a branch while I was crawling through some bushes. I didn't know it at the time but I think that necklace was a good luck charm."

Bills eyes rolled a bit, the motor-home veered towards the ditch, just before I was going to yell at him he brought it back between the white lines. I knew what he was thinking but plowed on with my narration.

"As I was saying, for six day we had glassed up to a dozen bucks a day. Big bucks and little bucks but all bucks just the same. Thane and I thought we'd finally had found mule deer paradise, it was only a matter of time before we ran across one of the resident monster bucks. On the seventh day, the day after I lost the necklace, zero deer, didn't even see a doe. Eighth day zip, ninth zilch. We had to head home so came down on the tenth day. So after losing my good luck charm on the sixth day we quit seeing deer, it's like they all disappeared."

"You about done" Bill said looking over at me.

"Ya, well what do you think of that" I said.

"I think that I don't know where to go and you are caught up with this good luck stuff" he huffed.

I guess that was my cue to quit yappin and start navigating. Shortly after we waddled our caravan down a narrow valley bottom road to where a small camp spot sat pleasantly on the side of a pristine lake. On both sides of the lake, blue marbled peaks rose towering through the heavy winter air. It was too late in the day to do anything except look after the horses and set up for the night. Bill and I were in the heart of west central BC; a couple inches of new snow blanketed the adjoining mountainsides. It was late November and we were hoping that the mule deer would be rutting. We planned to hunt the rolling plateaus on horseback and returning to the comfort of the motor home each evening. A logging road, still a few miles down the main road would get us a ways up the mountain side, from there I knew of a cow trail that led right up over the top. Passing time in the motor home that night we looked at the topo maps I had brought. On the back of one, a friend, who had given it to me years earlier, had written:

Dawson, look up in that valley marked in red, there's 32" wide bucks there, if anybody asks you where you shot your deer, tell em in the lungs, cheers... Mitch"

"Did you guys ever shoot any that big?" Bill asked

"Nope, but I saw one that big and we carted some sheds off that were that big" I replied

In the fresh mountain morning we moved our motor home up the old road, at the end we set it up and spent the day searching for the trail that led up through a hidden draw in the steep rocky face. The next day we stuffed lunches in our saddlebags and headed for the top. As we climbed higher the two inches of snow steadily got deeper, in a few drifted in spots the horses plowed through belly deep.

Once over the top I pointed out familiar landmarks to Bill, "we camped over there, saw lots of deer down that basin" I said, pointing here and there. A spectacular November sun glistened over the snow-encrusted landscape; bordered by incredible serrated mountains, capped by a brilliant blue sky, the visual impact was exquisite. In many spots we would stop to glass and end up mesmerized by the extraordinary beauty that lay before us. I've read many articles describing the natural beauty of the Chilcotin region as one unparalleled. Having traveled into most of the finest geographical landscapes BC has to offer I can attest that the description is fitting. Deer tracks covered the snow; they were everywhere. In every stand of tree's we found beds melted into the snow, dig holes revealed where deer had pawed for graze. We poked around that day glassing down into draws, across ravines, penetrating the stunted pine patches with our optics. While we saw a few deer and a couple decent bucks we didn't see anything to get excited about.

Later that night in the motor home, Bill and I discussed the next day. Fidgeting with my *new* good luck necklace, one my daughter had made for me the day after I returned from that trip years earlier, I said we should try over on the slopes of the southern valley. Pointing to the area circled in red on my topo map, I showed Bill where Thane and I had seen the biggest bucks, "in fact" I continued "I saw the biggest buck I've ever seen up there. That's also where I lost my good luck necklace, maybe we'll find it," I said.

I don't think I'm a superstitious kind of guy, however, neither do I think I'm not, paradoxical but true. I just get a feeling; sometimes they pan out other times they don't. Bill is decidedly NOT superstitious but he's been known to follow a hunch or two. By definition an intuitive feeling is pretty much the same thing as a hunch, hmmm better to just leave that one alone. I've had my moments and so has he, I trust his hunches and I know he trusts my feelings. So the next morning while tacking up our horses in the dark he said, "so you wanna take a look down south eh", "yep, I think that'd be a good idea" I replied. Without another word he swung up onto his horse Mr and led the way up the cow trail. From the look of the awakening day we were in for a weather change. Once above the forest cover, gusting winds grew stronger as the day evolved. We stopped to brew a coffee in a little copse of trees midmorning to plan our attack.

"I'm going to head up the topside of these tree's," I said to Bill "why don't you go along this side and we'll meet up down the valley. When Thane and I hunted up here we saw lots of deer on this slope".

"Good idea, see ya in a bit" he said and rode off.

I headed back along the trail for a while before heading up a small ridge. On one side I had a terrific view of a fading away draw while ahead the valley side gradually opened up as I made my way higher. I reminisced to myself about sitting behind my spotting scope on this very ridge on one of those earlier trips. The topside of the stand of stunted pines was undulated with the hillside so I could follow along without exposing myself for further than a couple hundred yards. I turned down the valley to meet Bill in the distance. Rutted trails in the snow told tales of where the resident deer had traveled along the edges, stepping off my horse to look closer they appeared to be at least a few days old. Coming directly off the crown of the bald mountain a wider trail led down. I headed over to intercept it and was surprised to see grizzly tracks in the snow. While they were a few days old as well I was amazed to see bear tracks. I had thought that any bear would be snuggly secluded away in a winter den this late in the year. Mounting up I angled to get back on the deer trail. The wind was increasing. I had traveled about half way down the valley when I saw a flash of movement in the tree's ahead. My horse, Star, nittered. Must be Bill and MR I thought.

Star stopped and stared ahead, I followed his lead and stared at the spot I had seen the movement. Slipping out of the saddle I looked through my binoculars. As the view came to life in my binoculars I saw a bedded deer. Curled up in a thick patch of pines a buck was napping the afternoon away out of the crisp wind. Through my bino's I could make out horns but the jumbled branches stopped me from seeing anything beyond his ears. He was laying down facing into the wind unaware of our presence.

Without the knee deep snow and incessant wind I would've probably sat down and waited the deer out but given the circumstances that wasn't an option. I couldn't head down into the trees and hope to creep up on him, so instead I jumped back in the saddle and moseyed in the general direction where he lay. When I had closed the distance to about a hundred yards he must of heard or seen something and slowly stood up. Slipping out of my saddle again I squatted under Star. The deer stepped closer to an opening and looked downhill. Through my binoculars I deciphered horns from branches. Counting points I was disappointed to see only three on the close side, they were tall and very wide but I could only make out three points. The area we were hunting was a four-point or better region, and, we were after big deer anyways. My horse whinnied again during a brief calm spot in the wind; the deer looked back and stepped towards us. "Whoa" I whispered to Star. With the deer looking straight on I saw that my initial count was wrong, the big buck had five points on one side and four on the other. Before I could judge him anymore the deer walked behind a clump of tree's heading toward us. I took the opportunity and jerked my rifle from the scabbard. Star nittered again as the deer ghosted through the trees. Hooking the reins over my foot I sat down in the deep snow. As I followed the buck's movements with my binoculars his horns grew and grew as I watched him come closer. At forty yards he stepped through a small clearing and stopped to look at us, standing broadside. The hill behind him fell away, with open air behind him there wasn't a question as to how big his horns were. Tall and wide they looked to be close to the thirty inch wide mark. He swung his head to look downhill.

Neither my horse nor the deer flinched when my 7mm cracked in the gusting wind. Initially I thought that I had missed but a growing crimson spot showed my shot was true. The buck slowly stepped out of sight over the slope. There wasn't any need to shoot again I knew he wouldn't go very far.

After a few minutes I rode up to the fallen monarch and tied my horse off; he was bigger than I thought, a true champion of champions. My first impression when I had first seen him was that he wasn't that big, now I saw why I had thought that. He was an old buck, his body size small for his horn size. Most likely in his last winter, his front teeth were worn away. One side of his rack was flattened a bit from having to lay his head sideways to feed. Fresh pine bark was ground in the burs where he had made rubs but he hadn't any marks from fighting on his face or shoulders. The real sign of regression was apparent in his horns. The burs were over six and three quarter inches around, however, the main beam had shrunk considerable measuring just five and a half inches.

Star was having a fit tethered to a tree so I left the deer and went to find Bill before he went down into the next draw. I had only rode a short distance up the hill when I saw him leading his horse down from the crest.

"See anything" he yelled as he got closer, "no not much" I said, "I saw a pretty good four by three" he told me, "just below here in the trees. We just came up, over there about twenty five yards away" he said pointing just passed where my deer lay."

"Come on down here" I said pointing down the slope, "I'll show where I lost that necklace my daughter made me" "Give it up" he said.

About ten yards downhill he saw the crimson trail in the snow and grinned. "Ah ha" he yelled. "You bugger". Tying his horse off we squatted beside the deer in the snow and marveled yet again over the huge buck. "That's a hell of a buck," he said. "Didn't you hear the shot," I asked. "Couldn't hear anything in the wind" he replied "and we were sneaking along looking for that deer, there was another with it but I never got a good look".

After taking a bunch of pictures we skinned out the deer. Hanging game bags over my saddle we loaded the quarters, head and hide and headed down to our waiting motor home, me leading Star while Bill prowled the trees ahead for deer.

We saw a couple more deer on the way down but not the numbers we'd expect to see with the amount of sign present. The next morning we again ventured up the cow trail. At the top we turned south and crept through the thickest cover but hadn't seen anything by midday. Cresting a slope Bill yelled over the wind "what's that" pointing down towards the valley floor. "Looks like a cabin" I yelled back. Riding down we found a small cabin nestled in stand of trees. From writing on the wall we read that it had just been built the year before. "Sure wasn't up here when Thane and I were" I offered. "Must be an outfitters cabin or maybe a line cabin for the rancher who ranges cows up here". It looked like a prefab cabin, "bet they chopper'd it up in pieces and put it together right here" Bill said. Hanging from a nail in the roof was one sleeping bag "we should bring a bit of grub and another bag up tomorrow and stay here for a couple days" Bill said "save a couple hour ride every morning".

After brewing up a pot of coffee we decided to head further southwest to a spot I knew to be a good glassing spot. A few hours later we got there but the wind by then was gusting so hard that it was impossible to sit and glass for very long. In the short time we did glass we saw one nice four point tending a doe lower in the trees and another deer laying a short distance away. On the way back we poked through all the tree patches, while there was sign everywhere again we didn't see the deer we should've. Later that night we decided that maybe the deer had moved lower down the mountain, "the next few days should tell us Bill" said. "

In the morning we stuffed as much in our pockets and saddlebags as we could, Bill tied a sleeping bag behind his saddle. We rode up to the cabin and made home for a few days, cut firewood for the tiny wood stove and had lunch. After hunting a cold windy afternoon away, a stream of smoke from the stovepipe greeted us as we came back to the cabin in the dwindling light. During the evening, and lasting all night long, a cold gale force wind blasted the high country. My sleeping bag, the one left hanging on a nail, was a summer bag, the wood stove had a burning time of two hours. In short, I froze all night long while my buddy Bill slept comfortably in his regular bag.

Thick wet snow greeted us in the early morning dusk. Tacking the horses up we set off for the far glassing spot in the falling snow. Again, gusting wind kept us from glassing that basin, we moved elsewhere but by late in the day a storm front had moved in and hunting seemed fruitless, between the wind and snow we could barely see fifty yards. Turning back to the cabin we grabbed our gear and headed back down the mountain. The heavy falling snow and the ever-present wind accompanied our decent; another trip up the mountain was not in the cards. While we were loading our gear up, I climbed the ladder to the box on the motor homes roof, "hey Bill" I yelled, "pass me those horns", "coming up" he said. As he was passing them I looked him in the eye and said "you never did answered me about believing in good luck things or not"!

Unfortunately his answer is unprintable...

A Conspiracy of Events

A Grizzly Bear Adventure

con·spir·a·cy [kuhn-spir-uh-see] noun,
plural con·spir·a·cies.

1. the act of conspiring.
2. a combination of persons for a secret, unlawful, or evil purpose
3. an agreement by two or more persons to commit a crime, fraud, or other wrongful act.
4. an evil, treacherous, or surreptitious plan formulated in secret by two or more persons; a plot.
5. any concurrence in action; and combination in bringing about a given result.

Merriam-Webster.com

I suspect by applying definition #5 from the list above, my hunting partner Ken Schultz and I; along with a huge black grizzly (we had named "Frothy"); another equally love stricken, and large, red boar who we didn't name; and a cinnamon coloured sow the two boars were courting; it could be said we were all engaged in a grand conspiracy to meet on a cold Wednesday evening in late May at about 6500 feet of elevation on a mountain top in northern British Columbia's rugged and remote wilderness. But for that to be true then all the happenstances of the previous week's events were also supplementary factors to the grand conspiracy of events unravelling before us.

Ken and I were hunkered down behind our backpacks on a bare, cold, north facing talus slope at 10:30 pm after a seven hour climb from the valley bottom below. Across the talus and avalanche slope on the opposite hillside two hundred yards away a big jet black and equally large ginger coloured grizzly were bedded down in the buck brush. The third bear was nowhere in sight.

Daylight was slowly losing its tug of war with the impeding darkness; long shadow's fell over the basin and a chill from the snow pack all around us was creeping in.

It was fantastic setting, for a fantastic event, for a fantastic conspiracy...

"Give it another try" I whispered to Ken

Ken mouthed his rabbit call and blew....

"*Rah Rah Rah Rah Rahhhhhhhhhh*" echoed through the basin.

The call was so out of place, so contrary to the tranquility of the space, it was like auditable graffiti.

"*Rah Rah Rah Rah Rahhhhhhhhh*" Ken blew his squealer again..

Through my binoculars I watched the two huge bears, curled up in the buck brush, lazily roll their heads and look our way; as if to say no big deal, nothing to get excited about.

As the rabbit screams faded away into the basin and were absorbed by the timber below I heard a loud;

"*Oorph – argg – Oorph*".

Ken said "holy crap there's another bear below coming in"

Almost at the same time as Ken pointed to where the third bear was coming from the two bedded bears got up and started coming towards us. I guess the third bears guttural grunting got them excited and they decided a meal of rabbit was a good idea after all.

The bottom bears "*Oorph – argg – Oorph*" grew louder as it came closer.

The top two bears joined the rush and were coming fast, it was last one to the table time!

At any rate and regardless of the whys and how comes. Ken and I were hunkered down on a rock slope in rapidly declining light with three grizzly bears charging across the hillside, grunting, popping and growling as they came. It was sooooooo cool. Ken looked up at me with a huge grin on his face and said "man, we need a video camera, this is great"

Ken, Geoff Helfrich and I were in Northern BC on a two week spring grizzly bear hunting trip. We had hunted these valleys, mountains and avalanche chutes before and knew that the snow pack and weather conditions were a huge part of any success, as was sitting and glassing the *"right spots"* of the mountain tops and slides as they green up.

The conspiracy to get to the moment at hand was a grand one. We had arrived to the valley a week or so earlier to find the snow conditions perfect, steep avalanche chutes were just starting to green up and the weather looked to favor our time. As a light rain chased daylight from the day we set a wall tent up beside a small river that runs down the valley. Once camp was set up we cooked a quick meal and hit our sleeping bags. The planned routine had us getting up at four am and start glassing right from camp. As the sun chased shadows from the dark north facing chutes we would move up the valley glassing avalanche slides and bare slopes as we went.

On the first morning, while water for coffee was warming up we set up our spotting scopes. Before long Ken spotted tracks in a high snow patch and soon after he found a bear way up on a ridge.

"Right on the edge of that slope" he said pointing about five miles away

Geoff and I zoomed in and sure enough, day one, morning one, and in the first hour we had located a lone grizzly bear feeding on a high slope. We watched as the bear fed, moved and fed, moved and fed then simply fed right over the crest line into another hidden basin. Spring bear, or at least hunting spring bears in the mountains, is a test of good optics, weather and patience. Over the previous seasons we had all "timed" the myriad of slopes, avalanche chutes and ridgelines we glassed: that one was a five hour climb, that one was a seven hour climb and so on. We had the timing down, now we just watched for a bear that would come into a slide or ridge and stay there. Spring bears are on the move, they are looking for food and a mate, not necessarily in that order. Over the next few days we continued our schedule of getting up early and following the sun up the valley as it pushed the dark cold shadows away, our day diary looked good:

> 1 large boar
> 3 single grizzlies, 2 smaller, 1-7' dark boar
> 4 single grizzlies, 1-7' boar, 1 – 8' boar 2 smaller boars
> 5 grizzlies, one big big black boar with a sow, single dark boar, 1 sow w/cub
> 3 grizzlies, 1 single boar 1 sow w/big cub

During the glassing sessions, as we sat in the warming sun behind our optics we were visited by noisy dickie birds, moths, butterflies and pesky mosquitos. Ducks and other water birds splashed and cavorted in the swollen river that bisected the valley floor. Spring in the north is a time of renew and everything was renewing around us. I was just boiling us a noon time coffee when I saw, without binoculars, three grizzlies on a snow patch high on a ridge we were watching. Earlier in the day Geoff had found tracks leading from a basin to the south so we had an inclination that the bears would show up on that ridge. Both Geoff and Ken saw them at the same time:

"I think that's that big black boar, Frothy" Ken said as we all got behind our spotting scopes.

We had seen that black boar a couple days earlier: Geoff and I had been climbing high on a ridgeline looking for a different bear and Ken was lower in the valley looking at a slide. Geoff and I were just above tree line looking into an avalanche chute when a sow came around the edge of the basin about eight hundred yards above us. Right behind her came this massive massive black grizzly. They stopped on the snow patch and he was just incredible. We didn't know it at the time but Ken has seen the two bears come along the ridge he was watching before dropping into where we saw them at. He was on the move trying to cut them off. Geoff and I looked at each other, "Holy crap" he said "that bear is a monster". I agreed, he was one of the biggest bears I had ever seen. The wind was good, it was about noon so we had lots of time left, things looked good, but, as with spring bear hunting, the bears fiddled around for a few minutes in the snow and then went back the way they came.

And it was Frothy, he was just as black and just as big as we thought, it appeared that him and another large red boar were pushing a cinnamon sow around. As we watched with our binoculars they chased her across the slope, then she would chase one of the boars back through the snow, snow flew as back and forth they went. Before long they had enough and moved higher to bed down in the snow bank, one boar on either side of the sow,

I looked at Ken, "that's about seven hours" I said "we better get going". Geoff stayed below to watch other chutes and slopes while we climbed through the forest and buck brush to the ridgeline above. Once there however, we couldn't find the bears. Their tracks in the snow went back and forth all over, the few green spots we could see were bare. We looked all over and into the adjoining basins to no avail. There was one trail of fresh tracks that led up and over a series of cliffs, we followed those and thought that they must have gone that way but there was no sense in following them. It was slowly getting dark when we looked at each other and said, "oh well lets head down". We were just crossing a talus slope to follow a slide down when Ken whispered, "Dawson, there they are"

Across the basin on the opposite hillside two hundred yards away we could make out the big black boar and equally large red grizzly bedded down in the buck brush. Ken and I looked around and quickly found a spot where we could get behind a couple rocks and hunker down. The sun was just creeping out of the basin when Kens first rabbit call;

"*Rah Rah Rah Rah Rahhhhhhhhhh*" echoed through the basin.

The call was so out of place, so contrary to the tranquility of the space, it was like auditable graffiti.

"*Rah Rah Rah Rah Rahhhhhhhhh*" Ken blew his rabbit squealer again.

Through my binoculars I watched the two huge bears, curled up in the brush, lazily roll their heads and look our way, as if to say no big deal, nothing to get excited about.

As the rabbit screams faded away into the basin and were absorbed by the timber below I heard a loud

"*Oorph – argg – Oorph*".

Ken said "there's another bear below coming in"

Almost at the same time as Ken pointed to where the third bear was the two bedded bears got up and started coming towards us. I guess the third bears guttural grunting got them excited and they decided a meal of rabbit was a good idea after all.

The bottom bears "*Oorph – argg – Oorph*" grew louder as it came closer.

The top two bears joined the rush and were coming fast, it was last one to the table time!

Crouched down on a rocky slope in rapidly declining light with three huge grizzly bears charging across the hillside grunting, popping and growling as they came. It was sooooooo cool. Ken looked up at me with a huge grin on his face and said "man we need a video camera, this is great"

Ken and I got ready, rifles over our backpacks as the bears came towards us, the lower bear was coming way faster than the top two. The hillside was undulating so we would see them coming then one or both or even all three would disappear for a second or two and then re appear.

But their noise told us where they were.

"There it is" Ken whispered as the lower bear came around the slope about forty yards below us. It was just on the other side of a little creek that came from the snow pack above. It stopped and looked towards the other bears, weighing his options. Just as he went to step our way a small gust brushed the back of my neck. Almost immediately the red bear startled to attention looking towards us, the top two bears noisy approach continued. The bottom bear swung his head back and forth, looking for the bears and looking for us, nostrils flaring as he tested the airwaves. Just as fast as the gust of wind came the bear turned and took off, huffing and grunting as he ran. Big bears look clumsy and waddly but when startled they are extremely fast, the red boar took off and was back in the treeline in seconds.

The other two bears kept coming across the slope, "ommphing" and "huffing" their way, oblivious to the retreat of the other bear. The black boar must have heard or sensed something because he shuffled downhill a few yards to look down from the small rise they were on, the red bear noticed this and stood full broadside about fifty yards away, looking at the black bear then back at where we were. Ken and I had decided earlier that if we found the black boar he would get first chance at it and I would get first chance at one of the red bears.

"Get ready Dawson" Ken whispered "I think we are going to get busted as soon as that red bear looks back"

The black boar was sensing something amiss in the air and turned to go back but kind of hung up. Even though he was less than sixty yards away we could just make out his back and top of his head.

"Better take that bear" Ken whispered as it too was getting antsy.

It was getting darker by the minute and the bears knew something was up but they had not yet seen or smelled us. I settled my crosshairs on the broadside bear and gentle squeezed the trigger. The big bear rocked under the impact and went down, the echo from my shot bounced back and forth in the basin. The big black bear swung around, looked at the other bear and took off, we caught brief glimpses of him as he went across the slope but he never offered Ken a shot.

We gave the bear a few minutes but it didn't matter, he was dead and probably was before he went down, a double lung and heart shot.

"Nice bear" I said, "man that was cool, what an adventure"

"Beautiful bear" Ken said as we marveled at the bear in the failing light.

In the dark, using headlamps, we set about skinning and deboning the bear. A few hours later we stepped off, with heavily loaded packs for a long dark, slippery decent. Somewhere around four thirty am, when we should be just getting up, we took our socks, boots and pants off for the cold river crossing, our last hurdle before getting back to camp.

Geoff was up and boiling coffee water when we dragged our cold tired butts into camp that morning. As another glorious day came to the valley Ken and I told Geoff how it all went down. It was, and is, a fantastic adventure that neither of us will soon forget.

Over the next week we saw a bunch more bears, both Geoff and Ken climbed after a bear or two but neither got an opportunity to shoot a bear. Each day spring evolved a bit more, the snow pack receded and was replaced by greenery, flocks of dickie birds flickered though the valley and the bears traveled farther and farther afield. Under a dark sky with rain filled clouds we broke camp and loaded for the twelve hour drive back home.

Our little valley had been a fantastic setting, for a fantastic couple weeks. Was it a conspiracy, who knows but it was a fantastic adventure...

Monarchs of the Peaks

A Goat Hunting Adventure

"Goats!"

"Where" I whispered

"Way over there"......

I followed my partner's finger point to a rock wall three drainage's away. I had been looking up a closer drainage but quickly swung the spotting scope around and focused on the distant white spots. There appeared to be a group of about twelve nannies and few kids scattered over the rocks. A big white dot caught my eye, one goat was bedded high above the group just below the crest. Even at the considerable distance I was looking from it appeared to be bigger than any in the group below.

"That one on top might be a Billie" I said

Pulling back from the spotting scope I said, "Man, that's a long.....long way off"

"Let's go get them" he replied.

"Maybe we should get a look at the other side of this mountain, we might be able to find some closer than them" I said gesturing to the far group of goats.

We ended up doing both. Going over the mountains crest to glass the other side, focusing our spotting scopes on distant alpine meadows and hour by hour we closed in on the goats spotted earlier.

This all was happening the first day of a late August goat trip in north western British Columbia. We had driven twelve hours from our home town of Prince George two days earlier, crossed a small lake and backpacked into this remote mountain range the day before. Northern BC in the fall is a grand place to be, mountain meadows are alive with color, high reaching peaks gently covered with the first dusting of winter snows. My partner for this trip was Stefano Amedao, owner of Rainbow Taxidermy in PG. I had moved to PG the year before and had met Stefano that spring. We had planned to hunt this range for seven days in hopes of finding *"Oreamnos Americanus "* monarchs of the peaks.

Probably the hardest thing to leave behind when I moved from Victoria to PG was my long time hunting partner, he and I had made many mountain backpack trips together. This was to be Stefano's first, he and I had discussed the rigors of backpacking and mountain hunting months before. Stefano had eagerly practiced with a loaded pack and was confident in his ability's for the adventure ahead. I was confident in his physical shape, but must admit to being a little apprehensive about his ability to handle the mental side of such a hunt. We each had personal goals for the days ahead. Stefano wanted to harvest a goat, any goat. I was hoping to find a large billie.

Leaving the truck parked beside the Cassier highway we stuffed last minute necessities into our packs, threw them in the boat and set off. Distant mountain tops beckoned us as we crossed the lake, the sun shone bright, the sky clear, a cool breeze whipped miniature whitecaps on the lakes surface. Weather is the biggest factor when backpacking, wet rainy weather is usually the norm, the sun was a welcomed surprise.

Beaching the boat, we dragged it across the rocks and stashed it in the bush. I snapped a few "before" pictures as we shouldered our packs and pushed into the forest. We settled into a nice casual pace. The first and flattest section was soon behind us. Stefano mentioned that this wasn't as hard as he imagined. I didn't have the heart to tell him that we had to get past this flat valley bottom before the real climbing started. We stopped often for rest breaks, gulps of water and to strip off layers of clothing. Up rock slopes we went, across buck brush infested chutes, through pockets of tree's, stop, pant, do it again. Hours later we stopped and had the first glimpse of the alpine meadows above. There was still plenty of climbing left but we were getting there.

Backpacking up mountains is mostly a state of mind. You definitely need to be in good shape, but even then it all boils down to a mental game. The human mind has amazing powers, tell yourself you can't do it and you won't be able to. Tear off for the top at a breakneck pace will only get you a quarter of the way up gasping for oxygen. Pre-trip preparation and a comfortable pack are a must. Trying to pack one hundred pounds of gadgetry and non-essentials is the biggest pitfall of all. I remember a climbing trip I went on years ago. One of the guys that came had packed his gear at two in the morning the day we left! Halfway up the valley he looked near dead. I played hockey with this guy so knew him to be in good shape, the pace we were traveling was fairly mild but he was exhausted. A quick check in his pack was enough. I was exhausted just looking at what he was carrying, canned food, a frying pan, air mattress, AIR MATTRESS. I couldn't believe it. We convinced him to ditch at least half of his packs contents (including the mattress). Needless to say the remainder of the trip was easier for him. The moral of this is to get in shape, expect to be worked hard and go light. If you absolutely need it take it. If it would be nice to have, leave it at home. I like my pack to weigh about fifty five pounds, no more. Expendable's account for about ten pounds of any pack. A successful hunt and you'll have at least one hundred and twenty five pounds to pack back! And more than once, tents sleeping bags and other gear have been left tied to a tree.

    A few hours later we dropped our packs just short of the tree line. After a quick look around we picked out a relatively flat spot in the midst of pine trees to pitch camp. A light rain started to fall just as we got the tent up. We wolfed a meal of instant mash and headed for the peaks above. It was almost dark but the urge to see what we came for was to strong. I often dream of breaking tree line and finding meadows full of game. I have yet to find this hunter's utopia but the promise of such a spot existing lures me back year after year. We didn't see anything that evening, but there was sign everywhere! Full of anticipation for the mornings adventure we descended back to camp, wet from the increasing rainfall but happy. Around the fire that night we offered plans to each other for the following day, scrapped them, came up with new ones and scrapped them. This went on until we climbed into the bags and continued for unknown hours after.

    Awakening to a fresh crisp mountain morning we gobbled some oats, grabbed our day packs and headed out. Sometime in the night it had stopped raining but the bush was still dripping with moisture. We were both soaked before reaching the alpine. Stopping at the tree line we glassed the crags above. Seeing nothing we moved to a vantage point found the night before. That's where Stefano spotted the distant goats.

    As we went over the crest fog rolled in and out, rain sputtered and then poured. We moved along when the fog rolled in, glassed far drainage's and the opposite valley when it rolled out. During one clear spell we spotted goats on the distant rocks, lots of goats. From one position we counted about thirty goats, groups as big as sixteen, singles, doubles, big and little. I was beginning to think that the utopia of my dreams had been found, but that utopia was a two days pack away. Sometimes dreams are better left as dreams, at least these one's were. The thick fog made progress hard to gage. Early afternoon found us sneaking over the fog shrouded crest. Spotting the goats below us but still two drainage away, we decided to go down into the next drainage and come up the other side. If all went well we would crest the far ridge about two hundred yards above them.

Marmot whistles marked our descent into the valley bottom. Fresh grizzly tracks pocked the sandy trail. Years of use by sheep, goats and bears had etched the trail permanently into the ground. Following the trail across a precarious shale slope we stopped to gather our breath. Looking across at the other side of the valley we spotted a group of stone sheep grazing high on the grassy slopes, out came the spotting scopes, all ewes. One last burst to the ridge, we stopped just before the crest and loaded our guns.

"They should be just below us" I whispered, and crawled to the edge.

We looked and peered but they were gone. "Where did" "What the" "Oh man" all came out in the same breath. The goats were know where in sight. A closer inspection revealed a series of fresh trails in the loose soil.

"Looks like they went over that saddle there" I said pointing to the trails.

"Let's head up over this ridge and see if they are on the other side of those rocks".

I continued on as Stefano stopped for a drink in a small seep. Half way across the open ridge top I glanced down, goats were spread out like dandelions in my yard. Apparently they had only fed over the crest into this drainage. Dropping to the ground I managed to get Stefano's attention as he came into view. We crawled in behind a knob and set about surveying the goats below. They were about two hundred yards distant, some sleeping, some nibbling the short grass. The big one we thought to be a Billie was about five hundred yards higher. None of them had a clue that their mountain home had been invaded by a couple of camo clad hunters. We quickly formulated a plan. Stefano was going to stay on this knob, I was going to try and get around and above the big one. To get where I wanted to be I was going to have to backtrack a bit, go up over the crest and across the open tundra into the next avalanche chute. All was going well until I was about half way across the slope. Looking backwards I spotted four more goats high on a ridge behind Stefano. He was hidden by some peaks, but I was clear out in the open tundra. If I could see them, then they could see me, the jig was up!

    The only way out was to stand up and walk straight away from them. The goats below were hidden by the slope of the land, if the high ones bolted I was sure they would spook the lower ones, but they didn't. I stole a peek at them as I slipped down into the chute. They had showed no interest in me at all. This just might come together I thought. A sheer wall of rock lay between the goat and me. Stashing my day pack I inched my way to a gap in the rock wall and slowly peeked over the ledge. He lay in his bed about forty yards away, I stole a hurried glance at the other goats. All were peacefully going about their business. Somehow that scramble across the tundra had worked.

    Quietly I pushed my rifle ahead and steadied the crosshairs just above the goat's front shoulder. Silently coaching myself, "breath in, relax, squeeze don't pull or jerk". My .270 sent the 140 grain Nosler ballistic tip on its way. As the report echoed down into the valley floor, goats were up and about all over the place. My goat was down, it hadn't moved except to roll backwards. I watched in amazement as a line of goats scaled a seemingly shear wall.

A while later we met at the downed goat, back-slapped, yehadd, oogled and ogled! A couple of the fleeing goats stopped on a ledge only two hundred yards away. We looked on as they finally slipped away.

Closer inspection revealed that my goat was a nanny, the horns measured a respectable nine inches. I was satisfied, while not the trophy I coveted, the hunt was more than expected. To get that close to a wary animal is what I like about hunting. The actual taking of game is a bonus to me, the exhilaration of getting close is my reward. I snapped a few pictures as we quickly dressed the goat. By then the fog had rolled in and it was raining real heavy. A long wet hike lay before us and it was getting late. We decided to leave the packing until the next day and turned for camp.

It was late in the night when two tired, hungry and soaked hunters crawled through the bush to camp that night. A fire was the first order of business, drying clothing was second, scarf dinner and into the bags was last. Just before I drifted off Stefano said that maybe mountain hunting wasn't for him. It had been a long day no doubt, but experience tells me that when mountain hunting, all days are long. I mentioned that I was sure he'd feel differently come morning.

I was right, a night's sleep and dry cloths renewed Stefano's vigor. Over coffee and oatmeal we decided to take the big packs and both go get the goat. Since I had shot the day before I left my rifle in camp. It was raining hard as we headed for the distant peak were my goat lay.

Along the way we looked for a goat for Stefano but the foul weather made glassing next to impossible. By the time we reached my goat the ridge tops were socked in pretty good. Finding the goat was hard enough. While de-boning and capeing we decided to look for a different way back to camp. Not wanting to chance getting turned around on the crest in thick fog, we opted to go down through the next drainage, then side hill over to our camp. Loading the meat and cape in the packs we pushed off. Scrambling down shale scree is a knee pounding nerve racking experience. One missed step and you'd be sent into a tumbling rolling death slide. The grass covered mountain meadows you see from a far are in fact shoulder high shin tangle. We pushed through that stuff for hours, fell, got up and fell again. Fresh grizzly tracks were everywhere. I kept thinking of the smell of fresh goat meat drifting from my pack, and my rifle leaning against a tree, miles away!

Every ridge we crossed was supposed to be the last one. At one rest stop (ridge number nine) I told Stefano that we might have to crash in the tree's for the night. "No Way" he said, but by ten o'clock that night I was done in. We had been soaked to the bone for hours, early stages of hypothermia had set in and we were both thoroughly exhausted. "Besides" I said "clamoring around avalanche chutes in the dark is a sure fire way to utilize your life insurance policy". With that I finally convinced him that we should head down while there was still light enough to find shelter. We slipped and slid down to the tree line and forged our way into a thick stand of pines. In no time we had a crackling fire going, clothesline up and fresh goat meat roasting on sticks. Stefano even managed a chuckle, here we were huddled around a camp fire in our underwear eating goat meat on sticks. Lashing winds and pounding rain buffeted our hidey hole but at least we had the comforts of a fire. "And you thought yesterday was tough" I said.

Dawn broke to a bright clear new day, the storm had blew over during the night. We dressed in half dry clothes riddled with burn marks from laying too close to the fire and headed out to find our camp. It was noon by the time we trudged the last few yards to the relative comfort of our tent, it had taken us twenty nine hours to retrieve the goat.

Replenishing tired muscles with rest and food we lolled around the remainder of the day. The next morning we decided that enough was enough. The previous three days had drained our mental and physical reserves. A goat for Stefano would have to wait until next year. We packed up and headed down the mountain to the waiting boat. As luck would have it our decent to the lake was under a cloudless sunny sky.

Season of the Wolf

A Wolf Hunting Adventure

The rumble of our snowmobiles echoed throughout the white stillness as we slowed to a stop on the frozen lake. Stooping over a trail of fresh prints etched into the pristine snow, my partner, Bill looked in the direction the trail headed, "Give it a try" he said

Winters snow lay deep as far as the eye could see, a few stars twinkled in the approaching darkness, cold shadows crept onto the lakes white mantle. Sucking in a big breath of icy air I cupped my hands and let go with a long drawn out howl.

"Ahaaaoooooooooooooooaaaaa, Ahaaaaooooooooooooaaaa, Ahoooooooooooo"

"Not bad" he said

Almost immediately an answering call from a wolf drifted to us from afar.

"Cool" I said

Another series of calls shattered the silence. "Ohhhhwoooooooooaaa Ohhhhwoooo"

TALES FROM THE TRAIL

The call of a wolf personifies the untouched wilderness like nothing else, mysterious and romantic at the same time, it springs forth with unbound passion. It's the primal voice of an untamed unaltered wild spirit, almost taunting mankind to try and make it convert to man's way. Hated by some, feared and revered by others, the northern timber wolf was what had drawn myself and Bill Cash to the remote country west of Prince George, British Columbia. Bill had invited me to join him on a wolf hunting trip to the north central area of B.C. in the middle of February, the wolves mating season. Before riding into the white wasteland we had stopped in to talk to a local rancher Bill knew.

"You back again Cashman" the rancher said, "Just can't get enough of chasing them wolves eh" he chuckled.

"Hey Barry how's it going, this is Dawson Smith" he said gesturing at me "seen any wolves up this way lately"

"Ya, I saw a pack of thirteen the other day down at the river crossing" he replied

"Wow, thirteen that's a good sized pack" I said

"Saw them as I was coming down a steep hillside or there wouldn't be thirteen left" he answered, his steely eye assuring the truth was spoken. "I lose a few head every winter to them buggers, I see em I shoot em"

We talked for a while as tendrils of steam drifted from our mouths in the chilly farm yard. Not wanting to leave the ride in till dark Bill broke up our friendly discussion, "we better get going before it gets late"

"You boys going down to that cabin on the far lake" the rancher asked

"Yup" Bill replied

"I might ride down in a few days to see how you guys are doing"

"OK, see you then" we said jumping in the warm pickup.

Arriving at our destination, we loaded a week's worth of gear and fuel in the skimmers, and prepared to snowmobile into a trapper's cabin at the far end of the lake. This was to be our base camp from where we would set out each day looking for fresh wolf sign. During the drive up Bill had explained how we would hunt "*Canis Lupus*". "We'll just jump from lake to lake along creeks and swamps" he said "hopefully we'll find fresh sign, then try to call them out on the ice".

Both Bill and I had shot wolves before, but not by calling them in. A friend of mine had told us how to call wolves, we were both skeptical but wanted to give it a try. Our skepticism vanished as the magical sound of the wolves drifted to our cold ears at the beginning of this story.

"They are over that way" Bill said pointing to a tree covered hillside.

Long shadows crept over the frozen land as first ours, then the wolves, calls bounced back and forth. Bill and I had followed them to the frozen lake. We had seen their spoor, tracks and kills, had stopped along the way calling but this was the first time they had answered our calls. Earlier in the day we had found where the pack had laid out on the ice, where the alpha female had marked the ground with her mating sign, and where the alpha male had been challenged for dominance. Late into the dark night we continued the enchanting exchange.

Wolves have a great advantage over their prey in the winter, big feet keep to them afloat in deep snow, a striding gait and unmatched stamina that they can keep up for hours on end. Their prey, deer and moose in the north, break through winters mantle of snow and flounder. A pack of wolves will chase them until exhausted, then move in for the kill. Usually, the prey tries to get out on the ice where they can maneuver better, but even on ice a pack of wolves can kill an animal that they wouldn't normally be able to without winters benefit.

Our first adventure ended on day three with a broken snowmobile, a grueling sixteen hour day cutting eight miles of trail and not a single track seen. A week later we were back.

During the week we were in town, a wolf pack had killed a deer on the ice directly in front of the cabin we had stayed in. The trail we had cut to the next lake was covered in wolf sign!

Encouraged by the abundant sign we headed out the next morning, surely we'd see some today I thought. Bill's knowledge of the waterways we traveled was unbelievable, we would stop on a lake, call, then move through a series of swamps to another.

Coming to the end of one lake he pointed down the hill and mentioned that he thought there was a lake below, before I could question him he took off. Weaving through the tree's for better than an hour we came out another frozen lake. With a big grin Bill told me he knew it was there somewhere. I could only return his smile "good thing" I said "getting back up the hill through the trees would've been almost impossible"

Somehow we traveled through the labyrinth of waterways without losing our bearings, searching the whole day but never finding wolf sign. Riding into camp that night I looked at the trip meter on my machine.

"just over a hundred kilometers" I said

"Know what you mean" Bill replied rubbing his butt, "that was a pretty good haul"

Heat from a roaring fire in the barrel stove chased the chills from our tired bodies. Other than not seeing wolves the day's travels had been great, we had seen moose, deer and coyotes, we had watched a pair of otters frolic on the ice beside an open spot in a river. Over dinner that night we tried to figure out where the pack could have gone. "What's that" Bill said perking an ear towards the door, the unmistakable whine of a snowmobile broke the evenings calm. "Must be Barry" I replied.

Looking down the lake we saw a single headlight beam pointing our way. Half an hour later the rancher pulled up to the cabin.

"How's it going guys" he asked mustache and eyebrows white with frost.

Bill told him that we had broke a machine the week before and only just returned that morning.

"Too bad" he said. "There was a pack down this end two days ago, I saw where they killed a cow moose over on the far side, but I think they moved on. Jim on the third lake over said he saw them yesterday. You might catch up with them by the big mountain to the west"

Protected from winters brutal cold night, warm and cozy inside the little log cabin, swapping tales, this is how the early adventurers must of wasted away long winter evenings I thought. Outside the mercury dipped to minus thirty five, thumping from the frozen lake and the occasional crack from a frozen tree filled the stillness. Our talk surrounded the routes we had ridden that day. The whole adventure seemed surreal, straight out of a Jack London or Lawrence Mott tale, I was enthralled.

The next morning Bill and I traveled a series of creeks and rivers to get to the mountain. Bill knew of another cabin we would move camp to. Not a single road dissected the pristine wilderness we rode in. The lakes we crossed are only accessible by float planes, the rivers by boat, the vast forests either snowmobiles in winter or hikers and horses the rest of the year. The sense of adventure was over powering. I must admit to feeling a bit like Jack London's fictional character Malamute Kid as we etched our trail deeper into the unmarred hinterland. I have backpacked into some of BC's most rugged mountains, climbed virgin peaks, explored remote valleys that few others have set foot in, but none of those adventures could equal the adventurous spirit that came over me as we traveled the white landscape. Preparation of such a trip is crucial, with temperatures as low as minus fifty northern BC in winter is not a place to go unprepared. Without Bill's knowledge of trapper's cabins I wouldn't think of getting that far back in the bush. Our skimmers held fuel, food, clothing and camping essentials. A good reliable snowmobile is critical. We had managed to limp Bill's machine out the week before, but we were lucky.

Arriving at the second camp just before nightfall, we ditched our skimmers and headed to a small lake, had just began calling when the wolves answered. We retreated to the cabin after the unique experience of engaging a wolf pack in a lengthily "discussion". Quickly, Bill had a roaring fire in the drum stove.

"That was awesome" I said

"Sure was, hopefully they'll stick around and answer our calls in the morning" Bill replied

"We should hide the sleds in the bush and get out on the ice", I said, "if they answer we'll crawl around on our hands and knees calling and stuff, that way we will look like other wolves, at least from a ways away"

"Sound good, are we boy dogs or girls", he said laughing

Our setup was perfect, we had arrived early, ditched the machines and got out to the middle of the lake. Except the wolves wouldn't answer. We called and called but to no avail. In the afternoon we found out why.

During the night the pack had killed a cow and calf moose, the chattering of ravens had alerted us to the kills. We stalked into the kill sight, but the wolves were far gone. A fresh wolf kill in winter is a sight to see, the purity of white snow obscenely marked with the contrasting bright red of blood, nature's version of graffiti. We recounted the kill from tracks and blood trails. I wondered aloud at the gruesome fierceness that wolves kill with, the moose had suffered greatly. The cow had fought her attackers feverishly, bleeding profusely, her entrails dragging in the snow, she had moved five hundred yards from the initial attack sight, turning on her attackers then retreating, ultimately falling in a stand of trees. Her calf suffered the same fate but only traveled one hundred yards or so. Other than one hind and some rib meat the cow lay uneaten, the calf had only killing marks, no meat was eaten from it. We had found numerous kills during our travels but this one seemed the most wasteful.

For three days we watched the kill sight closely. A wide variety of animals shared in the bounty, ravens, eagles, coyote's, mink and martins all came to the feast, but not the wolves.

"I think we should head farther into the mountains" Bill said over dinner that night

"Sure" I said, "they might be making a big circle"

The next few days we traveled across more lakes, down rivers, called, listened, traveling mile after mile without seeing sign of the pack. At one place we came to a river that was running. "I guess we have to head up stream to find a spot to cross" I said. Bill cracked a big grin "follow me" he yelled over the scream from his machine. Spinning around he throttled into the running water! I watched him head down river towards the shallow water. "Head for the ripples" Bill shouted from the other side "and don't slow down, you'll sink". I made it, but with far less precision than he had.

We finally located the pack again in the glow of a beautiful winters evening. Stopping on another small lake we called, an answering call came from deep in the dense forest. A few minutes later another rang out in the stillness, they were coming closer. We quickly took off our bright suits and headed out on the ice. Bill called while I paraded around like a dog, or tried to look like one. It was minus twenty eight degrees, we had stripped our heavy suits off and were flopping around on ice.

"We need a video of this" I whispered.

The wolves came to the lakes edge but were reluctant to come further. They would call and we would answer, exited yelps from the younger wolves answered every call we made. After an hour or so it was getting pretty cold crawling around on the ice. The sun had fallen below the horizon, we only had a few minutes of daylight left.

"I don't think there coming out" Bill whispered, "maybe we should get over to the next lake and see if they'll cross there"

Shivering, I said "let's go while we can still see" As fast as possible we got dressed and screamed for the next lake. A swamp lay between the two and we hit it going full out. Apparently the pack had the same idea. In the dwindling light I caught up as Bill slammed on the brakes and jumped from his machine, grabbing for his rifle. Out of the corner of my eye I saw a blond wolf running for the trees. Bailing off my machine I unslung my rifle. Shouldering it, I found the wolf in the scope centered the crosshairs and squeezed. The bark of my .270 pierced the night, in the falling light I saw the wolf fall in my scope. It was a young male, we marveled at the long legs and piercing eyes of the yearling wolf.

That was the last we saw or heard from the wolves, we looked for a few more days but couldn't find them. Somehow it didn't seem to matter that much. The night before we left, under a full moon, with stars shining like diamonds in the sky, the chorus of a wolf pack in full song floated over the still frozen night.

The hair on my neck stood on end, I wondered if the primal call was for their fallen brother or were they gathering for a hunt...

Cats-n-Dogs

A Cougar Hunting Adventure

Growing up on British Columbia's Vancouver Island, I cut my hunting teeth pursuing the elusive blacktail deer. Weekend after weekend we would slip into the old growth forest at first light and spend our days sneaking around as quietly and slowly as possible, searching for the little deer. If at any time during the day I would snap a branch or even step on a crunchy leaf I would sit down and wait for five to ten minutes. Stealth, quiet and slow movement was the key to being successful. Many a days as we snuck around we would encounter, or catch a glimps of a cougar as they went about doing the same thing, going after the same prey. I suspect they were far more successful than us however the lesson at hand was to move slowly, be quiet as a mouse and use stealth.

"Ready" Brandon asked

"Yup, let's go" I said

Deep in the forest above us three redbone hounds bayed and bawled, their chorus echoed throughout the still air. We stole a quick look at the handheld GPS and saw the straight line that all three dog collars left, the hounds were on a cat and we had to catch up.

I was with a buddy, Brandon Tames, a few miles north of Cranbrook, BC hunting cougar, it was three days after Christmas and we were on the second "run" of the day. I had driven down from my home town of Prince George, BC on Boxing Day. It had been a horrendous sixteen hour white knuckle drive through relentless falling snow and icy roads. But, in order for our hunt to be successful we needed fresh snow, so, the inclement weather was a good thing. Brandon is a young guy, but he grew up chasing cats (and bears) with his dad who is a fanatical hound hunter, as such he has far more cat hunting experience and knowledge than someone at twenty one should have. I had met Brandon a few years earlier through a mutual hunting friend and a hunt with his hounds was planned. Over the previous couple days our routine was to get up real early, Brandon is a firm believer in the early bird gets the worm, and head out to the bush roads looking for tracks in the fresh snow.

Of the estimated 4000 cats roaming wild in Canada there are approx. 3500 in British Columbia. Vancouver Island is reported to have the worlds, and BC's, highest concentration of *"felis concouler"*, the remaining BC population is spread throughout the lower half of the province. The east Kootenays, particularly up and down the Columbia River corridor, has long been a hot spot for hunting the tawny beasts.

Earlier that morning Brandon and I had come across a fresh set of tracks crossing the road, it was a few hours before light so we drove back around the mountain to see if the cat had kept going or was in between the roads somewhere. It was a cold dark morning, soft light snow landed on top of a fresh few inches of overnight snow, the conditions were perfect. The minutes ticked by like hours as we sat in the warm truck waiting for first light, occasionally one of the dogs in the box would bark or howl, they too were getting excited. As we waited and between sips of hot coffee Brandon showed me how his GPS collars and tracker worked, it is a neat piece of equipment. The dogs wore GPS collars and each dog represented a "curser" on his handheld display. Once we let the dogs loose on a track each one would muddle about looking for the freshest scent which told them the direction the animal went, he called it finding the hot end of the trail. As the dogs worked around and around it showed up as squiggly lines on his display. Once one of the dogs sorted the tracks out and led off, noisy and raucous, the others would join in the chase. That would show up as a straight track on Brandon's unit.

Waiting for daylight is one of the tests of patience we hunters continually go through. It is an exciting time as anticipation of the day's events is high, hopes and dreams of success run wild and loose through our minds.

As the day came to life we got ourselves ready for the chase. Back in northern BC I am a horse hunter, my partner and I take our horses far into the mountains off the beaten track, most of the time traveling just on moose or caribou trails. Horses are strong, amazing creatures, the power you feel sitting astride a good mountain horse as it navigates up, down, through swamps and across rivers is substantial. When we started unloading the dogs from the box in the back of the truck I felt the same sense of raw power. These were strong, determined, single minded animals, holding them back on their leash was like holding back their purpose for living. Brandon took the lead dog, Copper, over to the cougar tracks in the snow and got him on it, what a powerful dog.

"Go find the cat" Brandon yelled and unclipped him.

Copper started running back and forth bawling and barking searching for the hot end as we set the other two howling, lunging females loose. I stood in the fresh snow and listened as the dogs took off down the slope into the forest. The hounds filled the morning's air with a vibration of sound and movement, even though they were a couple hundred yards deep into the forest their presence was palpable. It was nothing at all like I was used to, there was no stealth, no quiet and certainly there was no slow movement. It was a loud, forceful, crazed and controlled mayhem, I loved it.

Brandon and I stood in the falling snow watching the GPS display as the three cursers drew together into a straight line and went further away. Just when we thought it was time to get packs on and follow them the lines on the display became a squiggly mess.

"Oh oh, they lost the track" Brandon said to me. "We'll just stay here and see what they do"

We could still hear barks and bawling but it was apparent from that and the display that the dogs had lost the trail. We set off into the forest to see if we could help them find it. As we neared where the dogs were the mantle of snow was broken from them running back and forth, we looked and looked but it proved to be impossible to sort out where the cougar had went. The day was young so we clipped the hounds on leashes and made our way back to the truck.

Later that day we were making our way down out of a drainage where we had spent the late morning and early afternoon in looking for sign. We had seen some tracks further up the valley but they were older so not worth running. As we came down past a little lake a fresh trail crossed the road in the snow. We had driven in on the road so there was only our truck tracks coming in and it had snowed all day so things were looking up. We stopped and got out, Copper barked from the box in the back.

Sure enough it was a cougar track, Brandon and I followed it a bit to determine direction and he was trying to sort out how old it was. It looked very fresh.

"Let's get them out" Brandon said.

He got Copper out of the box on a leash and went over to the trail, the big dog's nose was down and it was all Brandon could do to hold him back.

"Get the other two Dawson" he said. I went back and opened the door on their side of the dog box and the two females came roaring over to where Brandon and Copper were. He started up the trail with the big male on the leash trying to hold him back, once he was sure they were on the cougar's trail he let Copper go.

"Get the cat" Brandon yelled into the still air.

The dogs took off like a shot, baying, bawling and bucking, it was incredible to see, to hear and to feel them work. As the hounds took off we went to the

GPS, all three dogs were together, moving fast and in a straight line.

"They are on a hot trail" Brandon said "let's get ready to go"

I grabbed my day pack and rifle as he climbed up on the dog box to watch the GPS and listen for the dogs. It sounded like they were a quite a ways away and still moving. Their track got to the ridgeline above us and then turned right paralleling the road. We jumped in the truck and drove further along watching the display as we went. It appeared at one point the dogs lost the trail but quickly after a few minutes they were back on it and lined out again. We kept going and got to the top of a hill and parked the truck, it sounded like the hounds were coming along the ridge above and just behind us a bit. We could hear them coming and we were following them on the GPS.

One of the neat features about the collars the dogs had on were "look up" sensors, basically when the dog looked up it showed on the screen. Brandon could also tell if they treed something by their barks. I didn't know the difference but the second their noise went from predominately barking to deep, loud long bays I heard it and knew they had treed something. The display confirmed it.

"Ready" Brandon asked

"Yup, let's go" I said

We weren't stealthy, quiet or slow, we hiked up the hill fast and without any concern of making noise. The baying from the dogs led us right in to where they had the cat treed. Rounding a stand of young fir trees I saw the dogs worrying around the base of a big tree, Copper was bouncing on his hind legs baying and looking up at the cat. The Cougar wasn't very far up the tree, as soon as it saw Brandon and me come around the stand of trees it jumped out of the tree it was perched in and took off. But, with three wound up, noisy, excited redbone hounds hot on its tail it didn't go far before it went up another tree. This time however it went way up high and settled on a big branch looking down. The dogs were going nuts bouncing and trying jump and climb up the tree, baying and barking. It was a loud, pulsating unruly scene. I kept looking up at the cat as Brandon and I got leashes on the dogs and tied them off a few yards away. It was a weird moment for me, usually I would be crawling across a talus slope or forest floor trying to keep out of sight of my chosen prey. This, like the rest of the adventure was far from my norm. The report from my rifle echoed through the frozen forest, the dogs bucked and bawled on their leads, there wasn't any need for being sneaky or quiet.

A fine snow fell as Brandon and I set about skinning the beautiful cat. "That was so cool" I said to Brandon over the din from the dogs.

The next morning I pondered the hunt as I swung my truck north for the long drive home. I had hunted cougars before but not with dogs, I had killed cougars before but in a totally different hunting approach. I am not sure if one way was better than the other or if it even mattered but I was feeling very thankful and fortunate for having had such an incredible adventure.

Tales from the Trail Photo Album #2

"Although I deeply love oceans, deserts, and other wild landscapes, it is only mountains that beckon me with that sort of painful magnetic pull to walk deeper and deeper into their beauty."

Victoria Erikson

TALES FROM THE TRAIL

Dawson and his grizzly bear
"A Conspiracy of Events"

Bill and his wolf
"Ursus Horribilis" The Second Time

Ken and Geoff climbing an avalanche chute
"A Conspiracy of Events"

Brandon and his dogs with a treed cougar
"Cats-n-Dogs"

ADVENTURE STORIES

Dawson with his bull moose
"Of a Quest for Bulls"

TALES FROM THE TRAIL

Bill and Geoff with horses
"Of a Quest for Bulls"

Bill with his mule Mercedes
"Of a Quest for Bulls"

ADVENTURE STORIES

Dawson with his antelope
"Pronghorns in the Desert"

Bill with his mountain goat
"A Northern Adventure"

TALES FROM THE TRAIL

Dawson with Bill's mule Mercedes
"Of a Quest for Bulls"

Dawson with his mule deer buck
"Odocoileus Hemionud on Horseback"

ADVENTURE STORIES

Geoff up on ridge after a grizzly bear
"A Conspiracy of Events"

Dawson bringing horses back to camp
"Of a Quest for Bulls"

TALES FROM THE TRAIL

Bill on MR
"Odocoileus Hemionud on Horseback"

Dawson and Bill
"A Northern Adventure"

ADVENTURE STORIES

Dawson
"The Four year Quest"

Bill and Geoff with horses
"Of a Quest for Bulls"

TALES FROM THE TRAIL

Geoff with his bull caribou
"Of a Quest for Bulls"

ADVENTURE STORIES

Dawson with his bull moose
"Of a Quest for Bulls"

Bill with his antelope
"Pronghorns in the Desert"

TALES FROM THE TRAIL

Dawson, his horse Jett and pack horses with bull caribou, "Of a Quest for Bulls"

Little cabin half way in
"Of a Quest for Bulls"

ADVENTURE STORIES

Dawson with grizzly bear
"Ursus Horribilis" The Second Time

Bill and Geoff on horses
"Of a Quest for Bulls"

TALES FROM THE TRAIL

Dawson with mountain goat
"A Northern Adventure"

Bill in cabin
"Season of the Wolf"

ADVENTURE STORIES

Dawson on his horse Jett with
pack horses, "Of a Quest for Bulls"

The country we hunted
"Of a Quest for Bulls"

TALES FROM THE TRAIL

Ken at the ridge after a long climb
"A Conspiracy of Events"

Dawson with his mule deer
"The Four year Quest"

ADVENTURE STORIES

Bill, "Hard Bargain Valley"

Dawson with cougar
"Cats-n-Dogs"

TALES FROM THE TRAIL

Dawson with mountain goat
"Monarchs of the Peaks"

A little range cabin
"Odocoileus Hemionud on Horseback"

ADVENTURE STORIES

Dawson with packhorse Diesel,
Of a Quest for Bulls"

Bill "Of a Quest for Bulls"

The Four Year Quest

A Mule Deer Adventure

Thump ..Tha..Thump..Tha..Thump..Thump – my heart pounded, three hundred and fifty yards across the frozen oat field stood a huge non typical mule deer, even with the naked eye I could tell he was big. I stole a look through my binoculars to confirm what I saw. A small gasp escaped my frozen mouth as a silent point count reached fourteen. Man what a deer! I would of shot but couldn't, he was mixed in with a bunch of does and smaller bucks, a shot was next to impossible. I was huddled in a clump of frozen willows on the second day of a search for a real bruiser. A search that had started four years earlier...

Four Years Earlier

While planning for the upcoming hunting season, my partner Thane Davies and I decided to go on a backpack trip for mule deer. We both were experienced backpackers and both had been successful hunting deer. But, neither of us had dropped the hammer on a real mega buck. That was our goal, we had two main criteria for selecting hunting locations; first we wanted big mulies, secondly and probably the most important, we wanted to get where no one else was. Our research pointed us towards the high mountains west of Williams Lake, British Columbia. The first segment on our search for the holy-grail was played out on the high plateau stages of Mt. Waddington, B.C.'s highest and most formidable mountain:

The first year – September 10th

As morning rays broke over the high peaks of Mount Waddington we parked on the shores of Sopeye Lake, a small lake at the base of the mountain. Looking across the lake and up, then up some more, we first laid eyes on where we hoped roamed a record book deer. Without too much reservations we shouldered our packs and headed around the end of the lake. From air photo's we had decided to follow a small creek up through the dense lodge-pole pine forest. After clearing the thick lower slopes our trail led us up a near vertical ridge. Hour after hour we pulled and prodded, grunted and groaned. Thane had destroyed his knee in a motorcycle accident the spring before and was only six month out of a full leg cast. I wondered if it would sustain the pounding he was subjecting it too. Maybe we should have waited a year for his leg to strengthen up!

Dusk found us stopping to camp at the bottom of an avalanche chute. We were not even close to the top and worst of all, Thanes knee was feeling the effects of carrying sixty five pounds of gear up 6000 feet of steep mountain side. In the dwindling light I headed off to find water while Thane set up the small tent. After dinner of magic pantry and instant coffee we renewed our enthusiasm with talk of monster bucks waiting above. Dawns light found us heading to the top with day packs. We were a hopeful pair. Our first glimpse of the rolling plateaus came in the early afternoon. It was one of the prettiest sights I had ever seen. We looked up and down, sideways, here and there, stared through the glasses, scoured the ground for even a single hoof print but didn't see an animal or any sign of deer all day. We retreated down the mountain to our awaiting tent in pitch black. We gave it a good try again the next few days then headed down the following day, heads hung low. We hadn't seen a single track. Dejected but not defeated we talked during the long drive home of finding the mule deer mecca next season.

TALES FROM THE TRAIL

The second year

With another year to research for a "better" spot, we scoured air photos and hunting reports from the Ministry of Environment. I telephoned local hunters and better yet, made friends with the local Conservation Officer. During the drive up, Thane and I both regained our confidence as I re read (for the thousandth time) a letter I had received from the C.O.; "Hi Dawson, good to hear from you again, I circled two terrific spots on the air photo's you sent, that mountain is a super big buck area, there's some dandy bucks on the high ridges above tree line, you'll do great. Remember, when anyone asks where you shot your 35" buck, tell them "in the lungs", cheers Mitch!

Ha ha, man this is going to be great. We parked the car at the end of an old logging road. Confidence does things to a guy, this time the mountain didn't look so high, Thanes leg was in good shape, and we were in mountain busting shape. We shouldered our packs and headed off, hell even the pack in seemed easy! The only unknown factor that could get in our way was the weather!

We spent a whole week looking into pea soup thick fog and rain. Our rations ran out after day three, we shot and ate ptarmigan for breakfast, lunch and dinner. We got wet on the pack up and were still wet when we came down. We saw a few deer but not what we climbed so high to get. As some sort of consolation prize Thane found some awesome sheds. The deer of our dreams were there, but this time the weather beat us. Again the retreat of the mountain was done with our chins in the mud, the crack of uncertainty was gaining girth. But again we leapt that hurdle and decide that next September we'd be back, hoping for better weather.

The third year

While planning our falls schedule, I called our friendly C.O. again. He re-confirmed that there was some monster bucks in them hills, "it was the weather that beat you guys last year" he said. Armed with that, and the sheds we had found I managed to convince Thane that the third time is always lucky. So off we went again to our mountain, with even better plans and gear than the last trip. This time my pack held a vestibule for the tent so we could dry cloths over the little stove. We brought lighter and more food, and we brought a little shaker of spice for the ptarmigans. Thane parked the car at the same spot as the year before. After a quick coffee we shouldered the packs and headed up the same trail that we had dragged our chins down the year before. The weather was perfect and the forecast called for the same all week long. We found our camp spot from the last trip and set camp. There was a few hours of daylight left, so we headed for the top, nine hundred feet higher. But we never made it, through the spotting scope Thane zoomed in on mule deer in the high alpine meadows, it was hard to make out what they were in the fading light, but this segment was starting out like we hoped.

Reluctantly we descended to camp. Before light the next morning, we grabbed day packs and rifles and slipped up through the trees and up the shale slope to the top. It wasn't long before we were seeing deer. I snuck to within forty yards of an average four pointer, the largest mule deer I had seen since our sojourn for mega bucks began. It was hard, but I watched him walk away. Thane spotted a gigantic black bear, deer were on the ridges, on the timberline, seemingly everywhere. By midday the sun was beating down, the deer were gone to their beds, and we were grinning like cheshire cats. From a high vantage point I spotted a flicker just on the edge of the tree's below, a spindly three pointer quietly slept the early fall afternoon away. I was in heaven, we had finally found the spot, the weather was good, we were bound to find a big boy up here.

The morning of day four I ran into "mio-grande", my buck, the one I had dreamt about. As mornings rays broke the ridge top I was sneaking up through the trees above camp, Thane had decided to take the next ridge up so he was off to my right somewhere. Stopping to catch my breath and scan the open spots above I took off my day pack and laid it and my rifle down. Raising the binoculars I searched the open areas above. Just as I was getting ready to head off I heard a small twig snap. Looking over I watched an enormous buck emerge from the bush above, fifteen yards away! I blinked hard, then blinked again, he was still there! His huge five point rack still carried summers velvet, the long tines seemed to reach the tree tops. Never had I seen such an awesome display of headgear, I estimated he was about thirty two inches wide. My heart hammered, slowly gaining my wits I realized my rifle was on the ground. Sneaking a peek to where it lay, I almost fainted! I had dropped my day pack on top of it! There was no way to retrieve it without making some commotion. Slowly the realization of the moment hit me, three long years boiled down to a stupid mistake. The big deer walked around a bush, his huge rack towered menacingly above the small tree, I grabbed for my gun, he heard me, he bounded into the trees below, I cried.

Thane had heard the deer take off and whistled, I whistled back and we met higher up. After much crying by me we decided to leave the deer and hope he wasn't spooked out of the country. The next few mornings we saw large tracks leading across the shale slope but never saw him again, the close encounter must have put him into a strict nocturnal mode. I was inconsolable, the golden ring was within my grasp, I had dropped the ball on the one yard line, we lost again. The crack of uncertainty had grown too big, our confidence was shattered. Maybe we couldn't do it. Maybe we should lower our standards and take a smaller buck. The drive home that year was like a funeral procession. When Thane dropped me off at home I said "we gotta go back", he just looked at me and smiled. Light at the end of the tunnel.

The fourth year

The fourth year was a different deal. I moved from Victoria to Prince George, BC. Thane and I went sheep hunting our usual mule deer week. Other hunting trips stacked up and we couldn't get to our mule deer quest. We were not going to hunt mulies, "maybe a year off would be good" he said. Fortunately for me, a friend in PG agreed to go with me on a late November deer trip to the Peace country of north eastern BC.

During the drive up I told him about Thanes and my quest for the "buck of all bucks", each and every grueling step up god forsaken terrain, the fog, the rain, thunderstorms, and lastly, the beating we had taken. I told him I wasn't going to sell out. Even though Thane wasn't along I was still going to hold out for a real buster. He just shook his head "you guys are nuts", "I just want some venison for the freezer". How I envied him. A business associate of mine had a farm in the Fort St. John area and had gave us permission to hunt it. We arrived to a bitter north wind and several feet of new snow.

The weather on our first morning hunting was minus twenty six and snowing like crazy, we poked around all day long, the deer must have been holed up. The second day is where this story began.

I had spotted the group of deer in the morning's chilly haze. There was about thirty deer in the field, two or three real good bucks but the big non typical was by far the best. Quietly I made my way undetected to the field's edge. My buddy on this trip was supposed to be on the other side of this field, I watched as the big buck followed a doe across the field, his nose following her movements. Their direction would put him in range of my hidden friend. The buck cleared the doe, stood sideways and looked back at the other deer. What a sight, matching tendrils of mist streamed from his nostrils. I was wondering if my buddy had seen him.

As if to answer my silent question a deafening boom erupted from the bush................ Beside me!

What the .. who? I looked over to my left and my friend was standing beside a tree. I looked out at the deer in the field. There was a nice four point looking at him, the big buck stood looking as well.

Boom ....I couldn't figure out what he was shooting at, the big buck was still standing there. I snapped of the scope covers, leaned against tree and waited for the crosshairs to settle on his chest. The crack from my shot echoed in the frosty air, he didn't flinch at the shot. Quickly I chambered another round, raised the crosshairs above his back and squeezed. The 140 gr. nosler ballistic tip from my .270 found its mark, the huge deer fell over.

We waded through the knee high snow to where my deer lay, my heart pounding. What a magnificent beast, he was huge, his non typical rack held fourteen countable point grossing 194 1/2 BC. My four year ordeal was over.

A Northern Adventure

Horses, Mountain Goats and Muskeg

A small funnel of smoke rose gently into the still dark sky, the little lakes calm surface was broken by the familiar rings that revealed the presence of feeding trout. Six tired faces stared into the crackling flames, no one spoke. Bruised, muddied and exhausted from the day's travel, we sat quiet with our own thoughts. Quietly grazing in the moonlight on the shores edge were nine horses, they had carried our gear and us to the remote lake.

Quin's voice broke the tranquility...

"Don't get discouraged about horse hunting because of today you guys, that trail was defiantly not normal" he said.

His words started a flurry of conversation, the serene setting became loud with voices as we replayed the ride in, laughing and joking about the many tough spots we managed to get through.

I had been a whopper of a day on a whopper of a trail. Before long we all crawled into our bedrolls.......

The six of us had traveled fourteen hours from Prince George into the heart of north west British Columbia two days earlier. Myself, Bill Cash and his son Bill junior from PG, Bill Quin and Vern from the Kootenays, and my dad Ken Smith from Victoria. We planned on hunting the rugged Cassiar Mountains for stone sheep and mountain goats, Quin and Vern could only stay for a few days while the rest of us had ten days to hunt. Bill Cash and I had planned this hunt while wolf hunting the winter before, he had wanted to hunt stone sheep and mountain goat with his son. I wanted to take my dad on a mountain-hunting trip. Bill's friends Bill Quin and Vern jumped at the opportunity and made the 24-hour drive up with their horses.

We arrived at our destination, the end of a logging road, in the early morning light. Vern, Quin and my Dad looked in awe at the mountains as they slowly emerged from the morning's haze. Out came the spotting scopes as breakfast sizzled on a fire. Snow dusted ramparts rose high into the azure sky, mile high green meadows shone like emerald fields, north west BC in the fall is a glorious place to be, the colors as vibrant as an artist's palette. A friend of mine who lived in the area had told me about an old guide's trail that led into the remote mountain range we wanted to hunt. Unfortunately he wasn't sure where, or if, it met with the road, we had narrowed down the possible routes on a topo map. Our first day was spent trying to figure out where it lay in the vast land, finally deciding it must be off the end of a big burn.

The adventure begins

After breakfast on the second day we tacked up the horses, loaded the packhorses and prepared to leave. Somewhere in the dense forest across the burn was the trail my friend had told me about. It was supposed to lead between two lakes then wind towards the mountains base, once at the base it would follow a small river up into the valley. Getting across the burn was relatively easy until we hit the end of it. Blow down pine tree's lay crisscrossed like a child's game of scissor sticks.

"Okay boys" Bill said, we'll have to clear a trail through this mess.

As the others prepared a trail, Vern and I went ahead to figure out the easiest way to go, luckily, we found the trail half a mile through. Enthusiasm was high with only half a mile of trail to cut, axes flew, logs were heaved out of the way, a trail, albeit rude, was quickly made.

Vern and Junior led their horses through the maze, "It's not too bad" Junior yelled. Bill and Dad took their horses plus a packhorse through, all was going well until the sixth horse, Groucho started through. He decided that he didn't like our trail. The route led over one particular log then turned hard left, straight ahead was a couple of logs about four feet off the ground. Quin was leading the packed horse through the tangle. At the step over and turn left part he stalled. Quin pulled and pulled but he wasn't moving.

"Get on the other side of them blow downs Dawson" Quin said to me, "I'll pass you a halter rope, when I give you the go, pull as hard as you can". He was going to pull left as the horse moved, Vern readied to give Groucho a slap in the rear.

"Ready guy's" he yelled, "give er we replied"

Vern slapped, I pulled, Quin pulled. The startled packhorse reared back and jumped the blow downs. Straight into, or I mean, on, me!

All I saw was one very large brown object, an airborne object, coming right at me. I tried to get out of the way but it knocked me over backwards with its knees, landing on both my legs with its front feet. As I was rolling out of the way its rear hoof got me in the back.

"Dawson you oaky, holy christ" Quin hollered.

Panic ensued, Vern tried to gather the skitterish horse's halter ropes, Bill and Dad were hollering at me to get out of the way, my legs burned with pain, a heavy throb came from my back.

"Are you hurt" someone yelled, a loud moan was my reply. Quickly I got up and out of the way as the nervous horse stomped for footing. I slowly flexed my back to see if anything was broken. Pulling up my pants I saw that the horse's hoofs had only glanced off my shine, other than a little skin loss I was fine. Luckily I had on my daypack, it save my back from anything more serious than a bruise.

The trail started out great, but soon we got into muskeg. Not knowing what lay ahead we trudged on into the unknown. The unknown was muskeg, bog holes and swamps. Our horses got stuck in the mud time and time again, we had to lead them across an endless procession of mud holes. Bill and Quin were the experienced horsemen, we decided to let them lead the horses across the real bad swamps. Basically, they grabbed the horses halter rope and start running, someone would swat the horse in the butt as the runner took up the slack in the halter rope, and the race was on. If the puller tripped or fell, the lunging horse would trample him, it was that simple. To the inexperienced it looked downright dangerous, to the experienced it was still dangerous. I looked on in terror as a thousand pounds of lunging horse tried to catch up to a hundred and eighty pounds of running man.

"Hey Dad" I whispered, "there is no way anybody could get me to do that

"Me either" he said.

At mid-afternoon we took a break at a river crossing, lit a fire and had lunch while the horses rested and fed. Vern was starting to think we should turn back.

"This trail is the worst I have ever been on" he said, "my saddle was under mud on that last bog, I don't know if the horses can take much more"

"You got that right" Bill replied, "this is real bad, I didn't think we were going to get Groucho unstuck in the second hole back"

"Hey Dawson where's your topo map" someone asked

I pulled it out and tried to figure out how far we had come.

"We've come about ten miles so far" I said, "from what I can tell, the first small lake at the mountains base is only about five more miles"

"Well, we can't stay here" said Vern, "we either turn back or press on"

It was decide to press onto the little lake ahead. The next five miles were the toughest most grueling five miles any of us have ever gone. We crossed seventeen real bad swamps, the horses were stuck so many times we lost count. The last few miles the packhorses had to be unloaded before going across the bogs; we would pack the gear across and reload them on the other side. Darkness loomed as the lake shore came into view.

After a good sleep and breakfast we started up the valley trail.

Thick fog hung low as we left camp, the trail followed a river up into the valley. We had decided to leave the packhorses and gear behind at the lake, the plan was to ride to the top to see how bad the remaining trail would be. Except for a few boggy spots and a couple of river crossings, the trail was good. As we gained altitude the fog thinned, a light rain fell from the leaden sky. Quin and I broke out our spotting scopes at one of the crossings. High on an alpine meadow we looked on as a band of mountain goats feed in the mist. Junior's eyes grew large, he wanted to take off after them, assuring him that we would find others we mounted up and headed farther up the trail. Slowly, clear sky broke through the misty fog. At every opportunity we stopped and glassed. What a magical valley; draws, peaks, and slopes as far as you could see. This was the first trip into the north west mountains for Quin, Vern and my Dad, they eagerly soaked up the splendor before us as the fog lifted. We found the top camp at noon, the trail in hadn't been that bad, there was water and food for the horses, it looked good.

"Let's head back down to the lake and get our gear" Bill said. "Okay" came the reply.

Vern, Quin and I trailed behind as the others went ahead, we wanted to glass on the trip down. There was too much too see, we pointed out draws to each other, whispered what's that and soon had quietly wasted away an hour.

"We better get going and catch up to the others" Vern said.

The trail led down a steep hillside before crossing the river, climbing out of the river we came up over a knob and saw Bill standing on the other side of a bog hole, his stance signaled trouble.

"Uh oh" Vern said, "something's up"

After crossing the bog Bill explained the situation to us. Juniors horse, Star, had gotten Joe poked coming through the muddy hole. The stick had went up through his penis, scrotum and into his thigh. The poor horse was bleeding a water facet stream of blood.

"I don't know, but it's pretty bad" Bill explained, "I tried to stem the flow but it looks hopeless".

Quin, Vern and I walked up to have a look. Quin probed the wound with is finger and announced the same bad news, "It's bad Bill, I think it might have clipped an artery".

Shock started too set in as we looked on.

"I don't know what to do" Bill said

"There's not allot we can do" Quin replied.

My dad couldn't watch on as the horse slowly bleed to death, so he walked off down the trail. Junior was really upset, we all told him it wasn't his fault but he still felt responsible. It was a hopeless situation, the bleeding just wouldn't quit.

Six very concerned and upset hunters could only sit and watch as Star bleed in unimaginable quantities.

"He's bleed at least five gallons of blood" someone said.

We were sure that he was doomed, Bill thought he had better get Star off the trail before he died. He and I gently coaxed the ailing beast to walk down the trail, as he walked someone noticed that the bleeding slowed..

"Hmmm" Bill said "let's keep him going"

Giving Star a break at the river we discussed the situation. Even though he was doing well, the severity of the wound was so bad that no one thought he could survive the trip off the mountain. Bill opted to see if he would make the lake and decide what to do then,

"But he isn't going to make it through the bogs on the way out" Vern said.

"You guys go ahead" Bill said "let just get him to the lake"

In a light rain, Bill, Junior and I coaxed Star down the trail, the courageous horse moaned every time he had to jump something or flail through the mud. One of the hardest things I've ever had to do was to slap that horse with a twig to get him through a bog, but it had to be done. We rested the horse often, letting him set the pace. Darkness found us miles from the lake; wet exhausted and tired we pushed on,

"If Star can make it, we can" Junior said.

It turned out that we didn't take too much longer than the others guy's.

"You guys made good time" Dad said, "we just got back, we had a bit of fun on the way down".

Quin replayed the story over a cup of coffee, he was leading my horse Lightning, at one of the river crossings Lightning bolted out of line, Dad's horse Dakota decided to follow him, in the ensuing rodeo my Dad was thrown over the horses head. Quin chuckled

"I wasn't sure what to do, I thought Ken was dead, Dakota was stuck up to its neck in a mud hole, Lightning took off down the trail, what a mess" he laughed.

Our situation didn't look good, Vern and Quin were riding out the next day, they had to be home in two days. They were going to take Groucho out with them.

"What'll we do about Star" Dad asked.

Before turning in Bill decided he would ride out in the morning and get some medicine and oats for Star.

"That's gotta be thirty some odd miles through the mud" Junior said.

"It's either that or shoot him" Bill replied

We all bid Bill and his horse Mr good luck as they headed out the next morning. Vern and Quin left a few hours later. My Dad, Junior and I spent the day tending to Star and the other horses, dwindling the day away. Junior was getting worried when darkness loomed over the valley

"He should be back by now" he kept saying.

"I'll get up at first light and go find him" I said, secretly hoping that it wouldn't be necessary.

I couldn't shake the image of Bill leading that big horse through the bogs. If he's down I thought, then he's hurt real bad.

"Maybe I should head out and find him now" I quietly said to my Dad.

Just as darkness swept into the valley we heard a whinny come from the forest. Dakota, Mr's stable mate replied by the lakes edge, "that sounds like them now" Dad said. A huge sigh of relief escaped us all when Bill and Mr into camp.

The exhausted rider sat down as we tacked off Mr. Dad fixed Bill a coffee and some grub.

"I hate to say it" I said, "but you look like hell".

He and his horse were covered with mud and sweat.

"Hell and back" he muttered. "Mr didn't want to cross the holes on the way back in, I've been wrestling with that horse for hours, oh what a trail, I called a vet on my truck phone, she said there's nothing we can do for Star. I brought back some oats, a mineral block and butte to help with his pain, it'll work on the swelling too, but that's all we can do for him. Now the bad news"

"Bad news, what could be bad news" Junior said.

"Groucho didn't make it out" he said

"Oh jeez" Dad said "this trip is turning into a nightmare"

"It's not that bad" Bill said "he got stuck in a bog on the way out and Vern and Quin barely got him

Out. He laid down once they got him free and wouldn't get up, Quin told me he would've shot him but it wasn't his horse"

"Is he still down" Junior asked

"Nope, he was up when I got to him, but he doesn't look to good, I'm not sure he'll make it through the night, I left him some oats, he's between a couple bogs"

"How far back" my Dad asked

"About halfway to the beaver pond, Bill said, if we could get him to the pond he could graze in that field for days. But that's about ten bogs away from where he's at now, plus his saddle is in the bush were they had to cut it off him"

There wasn't any question of what we had to do, at daylight the next morning Bill gave Star his butte and we grabbed our packs and set out to rescue Groucho. A silent foursome hiked out the trail that morning, I'm sure all our minds were busy hoping Groucho was still alive.

He was, but he wasn't in very good shape. After feeding him some oats we had brought, Bill led him through the mud towards the beaver dam. I guess we had lady luck with us because Groucho made it without too much difficulty. Doing our best to corral Groucho in the pasture we headed back to our camp. Another nightfall found us trudging back to the lake, wet muddy and tired.

"Were not getting much hunting done" Bill said

"Oh well" Dad said "at least all the horses are still alive"

Bill chuckled "you got a point there Ken"

Over dinner we decided that with a sick horse that needed medicine, we couldn't move to the top camp

spot. Dad offered to stay at the lake while Bill Junior and I headed up the valley to hunt the next day. I would take my daypack with enough grub to hold us over for a day if we got stuck out.

The three of us rode out early the next morning. We stopped to glass draws, basins and slopes, saw goats and sheep but no big billies or rams. Just past noon we spotted what looked to be a good billie. He was feeding across a rock face.

"Let's wait and see where he goes" I said.

Swing the spotting scope I spotted four sheep high on an alpine pasture, ewes. Looking back we continued tracking the big billie as he fed across a saddle. Farther ahead on his trail I spotted what looked like two more goats.

"I think that's two more up on the edge of that glacier" I said pointing up a draw.

Junior looked but thought it was snow.

"No way" I said staring through my scope, "one of those snow piles just moved". The first billie fed right up to the others and lay down.

"We gotta get closer for a better look" I said.

"Let's ride up to the top of that tree line" Bill pointed.

From there we could tell that they were all billies, the two I could clearly see were big. It was going to be hard to get to them though, they were bedded at the tip of a scree slope that jutted into a basin. Their beds offered an almost 360 degree view.

"The only way we can get them is to go up that runoff channel" I gestured to the slope above us, "over the saddle, across that slope, over that peak and we should come out right above them".

"Let's go" Bill answered

We quickly unsaddled the horses and tied them to trees, I took my day pack and gave Junior my meat

pack. Up we climbed, I kept sneaking to the ridgeline to gage our progress, each time I peeked over the goats were in plain view. We climbed right to the top before a small rock outcropping shielded our presence from them. A small flock of ptarmigan burst from our feet, luckily they flew away from the goats. Over the peak we climbed, through the saddle, across a steep and precarious shale slope and up through a set of crags to the top. Finally we got on the ramparts directly above the bedded goats.

"Okay guy's" Bill whispered over the raging wind, "They should be down over that point"

Sneaking a peak over the rocks into the shale below I couldn't see them. Bill snuck over the crest into the basin but couldn't see them either. I had just started to climb over a series of bluffs when marmot whistles got my attention. Looking back I signaled for them to look into the bowl, nope Bill signaled. I crawled back onto the ridge top, looking down I saw all three billies crossing the slope below me. I waved but they had seen them already.

Lying down in the alpine grass I lay my rifle over the edge, they were a hundred yards or so straight below me. Finding the last goat in line, my scope steadied on its ribcage, gently I squeezed the trigger on my .30-06.

The report echoed endlessly in the high basins, the 150 gr. Nosler Ballistic tip found its mark. My goat fell on impact. Looking at the others I saw Bill shoot and his goat fell as well. Junior shot at the third goat but missed. Later we found that his scope was way off, it must have gotten banged in the scabbard.

I knew the billie I had shot was big but as I slid down the shale towards it my eyes grew and grew. It was huge. Bills jaw dropped as he and Junior came up to it "wow that thing is huge, did you see the other two" he asked.

"Yup" I replied "I think yours is bigger than this one"

"Really" he said pointing at mine "man what a goat"

We sat down and enjoyed the moment, the climb, the scenery, a successful stalk. This was Juniors first mountain hunt, he asked if all goats were this hard to get "not always" I replied "but I haven't found any of the easy ones yet" Bill chuckled.

We snapped off a bunch of pictures and set about capping and de-boning my goat. Loading the packs with meat, cape and horns we headed over to where Bills goat lay. Bills grin grew a bit as we compared the two, his was bigger, not by much, but bigger. I laid the tape to both goats, mine scored 51 1/2", Bills 52" even. We marveled at the good luck in finding three huge billies together. It was getting late in the day, we quickly capped out and de-boned the second billie.

"What's the best way down" Junior asked.

I said we should follow the glacier straight down until we hit the trail. Loaded with heavy packs we slipped and slid down the mountain. It was dark by the time we found the horses. Bill and I decide to head for the top camp as it was only a few miles away, he was going to ride one horse and we would pack the other two. Junior and I would hike up.

Junior and I finally walked into camp around two am that night, wet and cold. We had to cross the river in the dark, twice! Bill had the horses looked after so we just rolled into our sleeping bags.

The next morning we packed up and went back to the lake.

Resting up for a day, we geared up the following morning and headed out. Having two horses that couldn't carry anything we distributed the weight over the remaining horses and everybody walked. Groucho had benefited from three days at the beaver dam, he looked good. Slowly we eased the horses through the bogs towards the waiting pickups. We had a few problems, Dakota and MR both got stuck real badly, but the horses seemed to know that this was the way home and co-operated better than we had thought. A few yehaa's were yelled when the trucks came into view, we had made it, better than that all the horses made it. Bill Quin and Vern had told Bill that neither Star nor Groucho would make it back out to the trucks, we took a picture of Bill and I with our goats, Star and Groucho.

Dear diary:

Home from grizzly hunting, not sure our plan was as fool proof as we thought. The trip in was one of the hardest forty clicks we've ever traveled! It sure was cold up north in early May. Lost seven pounds, broke Bills quad, rolled mine, broke the skidoo, the skimmer, and my rifle. But boy howdy, did I ever end up shooting a booner of a bear, Bill got a beautiful wolf. He wants to apply for the same tags again next year. Hmmm, it wasn't really that cold now that I think about it, and now we know the way in past all that snow. Come to think of it, that trip for goats a couple years ago where we had all the troubles with the horses in the mud was way tougher than this one...

"Ursus Horribilis"
The Second Time

A Grizzly Bear Hunting Adventure

The Adventures of Tom Sawyer was the first novel ever to be written using a typewriter. The manuscript for the book was typed out on a Remington typewriter in 1875 by Mark Twain himself. Twain, however, wished to withhold that fact. He did not want to write testimonials, he said, or answer questions concerning the operation of that "newfangled thing", a typewriter.

From all accounts that we could gather, my partner Bill Cash and I were the first to access, or try to, a remote set of mountains in the early spring to hunt grizzly bears. The problem, we were told, numerous times, by numerous people, was the deep snow that plagues the area long into the spring season. One of the local outfitters went as far as telling us that there wasn't a snowball's chance in hell that we would get to the mountains that early in the year. Granted it would be easy to go all the way in on skidoo's, but skidoo's are not allowed in the area where we were heading. To get to our area we had to travel about forty kilometers off the Cassiar highway. An old roadbed would get us nineteen kilometers closer but the rest was up to us to figure out. Bill and I had hunted the area before on horseback so had a pretty good idea of what we were in for. Over the last few months, after we had received our LEH permits, we had come up with a plan!

Looking over at Bill as he labored in the rotten snow I wondered if we really did know what we were doing, and just how sound our carefully scripted plan was.

We had arrived two days earlier and had found the roadbed completely snowed in. The road, which should have gotten us up on a snow free plateau, instead, became the biggest barrier between our chosen start off point and us. Our quads, which were supposed to get us to the end of the roadbed, floundered and sat useless in the knee-deep snow, luckily we had brought a skidoo and a skimmer with us.

What we ended up doing was to slip the skimmer under the quads, take the winch line from the front of the quad around under the skimmer and suck it up tight under the quad, hook the skimmer up to the skidoo and "skiquaded" our way into never never land.

It all started out innocently enough when Bill and I sent in LEH application for spring grizzly bear tags in the harsh north-west corner of BC. A few months later we both received congratulatory letters from the Environment Ministry with grizzly bear authorizations attached. The year before, we had applied and were drawn to hunt the same area, however, everybody we spoke to said that we couldn't get the forty odd clicks off the road in spring. We had decided to heed others advise that time and chose to hunt in the fall only. While we saw some bears that fall we hadn't seen any of the monster bears we knew to inhabit the area. So a year later, wearing our best devil may care grins, we pulled out of our home town of Prince George with a one ton pickup, two quads, one quad trailer, one skidoo, one skimmer and a skidoo trailer hooked on behind for good luck. Twenty hours later we arrived at our destination to find the roadbed completely snowed in. That kind of took the lead out of our pencils a bit, we had counted on the first nineteen clicks or so to be the easy part.

"Well Dawson" Bill said "looks like we are going to put that idea to test a little ahead of schedule",

"Oh ya" I said shivering in the cold night, "the plan!"

"Give er" I yelled over the two stroke motors whine. Bill hammered the snow machines throttle and off we went. Slowly the track dug into the snow, momentum built, faster, faster and faster we went. Then we would bog down in rotten snow and lose momentum, just when it appeared that the skidoo would stall we'd hit a patch of tighter snow and our bizarre convoy would rocket off again. It was like the little red engine that could, only on steroids.

I think I can ..I think I can ..I think I cannnnnn, "Hold on Dawson" Bill would yell.

For the previous two days we had wrestled the machines and rotten snow uphill after hill, the last one before the plateau stood before us, taunting us to beat it, strong behind its defense of four feet of thick rotten snow. Our motley menagerie was careening up it with abandon, that devil may care swagger was back. Bill looked like a wild man in some low budget action film throwing his body from side to side on the skidoo, our camping gear was thrown loose over the snow for a few miles back. I was sitting on the quad holding on for dear life, it's motor thumping for all it was worth, transmission in fifth gear, the spinning tires helped the little bit needed to beat the snow, the top of the hill loomed closer, it's mettle sagging.

"Yehaaaa" Bill yelled, fist pumping the air as we broached the hill, his holler echoed down the valley. Before us was the vast expanse of the valley, except for a few shady patches where the snow was deep it looked like we were home free. Unloading the first quad we turned back down the trail to get the other one. Just before heading back we scanned the closer hillsides with our binoculars and were pleasantly surprised to see a black bear working its way down towards a small acre sized patch of burgeoning greenery. Bill and I exchanged a "new it" look and headed back down the trail.

We would spend another night in the truck and hopefully get the other quad in the following morning. To take advantage of the early mornings cold we opted to hit the trail at three am, "with any luck" I said "the snow will stiffen up and it'll be a breeze", famous last words!

Getting the second quad in turned out to be much the same as the first one but we had a taste of success, a little rotten snow wasn't about to stop us.

As dusk swept coldly over the vast valley I set up our small tent as Bill got a fire going, although we were three days into the hunt, or perhaps I should say adventure, this would be our first night away from the warm pick up. After a quick dinner we piled into our bags and froze to sleep.

The next morning we slurped coffee over our frozen lips as dawn reveled the splendor before us, the valley floor was almost bare of snow, however, the mountain slopes we were heading for lay deep within winters robe. The day coming to life promised to be a good one, as the cobalt blue sky slowly brightened, fingers of warm sunlight crept towards us over the eastern mountain peaks. We left camp and headed further inland on our quads, most of the way was relatively easy, a few deeper patches of snow presented small challenges that day but we steadily closed the distance to the mountains. We would stop on small rises in the terrain and glass as far as we could see, dip down into ravines, get stuck in the snow, winch ourselves out and climb another promenade from where we could glass again. We didn't see anything but the fact that we were hunting opposed to wrestling machines in the snow renewed the eternal optimistic attitude that is prevalent with hunters. Getting back to camp under the dying embers of a glorious spring sky we struck a fire and made dinner as long cold shadows pushed the day's warmth from the valley.

Leaving camp the next morning we headed for higher ground from where we could glass. The final installment of our plan was to be the easiest; at least that's what we had thought. A little hill protruded up from the valley floor, we were going to sit on it and glass the far southern slopes and avalanche chutes for bears. We had rode our horses up on it the previous fall and decided that would be the best way to locate bears on the vast slopes, unfortunately we forgot to factor a creek crossing and how much higher the water would be in the spring. Not to be persuaded by something as minor as that, we pulled out our axes and quickly made a bridge. Once across that we again dived into some deep snow. But aside from the mornings trivial travel problems we reached the high spot and spent a warm spring afternoon glassing for bears. We didn't see any bears that afternoon, what we did see however, was that the lower part of the avalanche chutes had shucked their winter coat and green was starting to show. We were certain that any bears coming out of their dens would show up in one of them in the near future.

The next few days were mirrors of that day, we would arise early in the morning chill grab a fast breakfast and head out for our glassing spot, the suns warmth would penetrate the valley by mid-morning and we would trek farther across the floor towards the distant avalanche chutes searching for bears or even bear sign. Aside from the occasional moose our only company seemed to be the abundant "dickey" birds that arrived in flocks returning from their winters sun-belt travels.

One morning just as we were setting off Bill whispered "Dawson, stay still".

"Ya right" I replied spinning slowly around. Bill was about ten yards from the tent and had grabbed his rifle, following the direction it was pointed I saw a gray/blond wolf crossing the burn. The wolf kept coming, seemingly oblivious to our presence, about sixty yards from our tent it paused a beat, looked towards us and turned broadside. Bill didn't waste a second, the echo from his rifle report bounced endlessly through the still morning.

"Now that's a good way to start off the day" Bill said as we skinned out the female wolf.

Later that morning, while lounging in the increasing warmth atop our lookout, we were talking about changing to another lookout when I spotted a dark spot moving across a snow-covered slope a few miles off. Snowbirds danced in the light breeze, somewhere in the timber below a woodpecker tapped outs in noisy feeding ritual.

"There" I whispered to Bill pointing to the far slope. Bill spun around training his binoculars on the hillside a couple miles away. Even with the naked eye we could make out the dark mass of a bear moving across the snow. Setting up our spotting scopes we peered closer. Rising heat waves blurred our spotting scopes as the big bruin worked down the slope heading for a slide area below.

"That's a pretty big bear", I said.

We quickly grabbed daypacks and set out to intercept the bear. As soon as we hoofed it off the ridge we would not be able to see him so we marked the slide by a distinguishing looking peak above it. We slipped and slid in the snow coming down off our lookout, once at the bottom we'd have to cross through a timber section that lay between two little lakes and then we would again be able to see up the slope, and hopefully the bear.

A couple hours or so later Bill and I peaked through the trees. Above us loomed the avalanche chute where we had spotted the bear. However, the bear was not in the slide as we thought he would be, after a quick you go that way I'll go this way discussion, Bill faded back into the tree's and headed south.

Loading my rifle I started north, it was a coin toss who would encounter the bear, if either of us would.

The early afternoons sun glared off the snowy slopes, shards of frozen ice glistened in the sun as they were carried aloft in the gentle wind high above, an occasional "phulmp" broke the cold stillness as patches of snow fell off tree branches, landing noisily on the snow laden ground. I quietly hoped that if I encountered the bruin he would be high on the slope ahead, a long way ahead. Hunting an animal as formidable as a grizzly bear I wanted time and space to plan out an approach, preferably undetected. After a few minutes I heard a crunching in the snow behind me and nervously looked back, Bill was following my footsteps. Once he caught up he told me that the next chute was wide open and there wasn't a track anywhere to be found.

"The bear had to have gone this way" he whispered. We slowly crept closer to an opening in the tree line.

The southern sun's presence showed itself in the chute, close to five hundred feet up the next basin was near snow free, wind-blown and avalanche uprooted trees lay crisscrossed and clear of snow on the lower slope. We stopped to glass further up the open sections only to be dis-heartened, the bear wasn't to be seen.

"Must of went around that next ridgeline" Bill said, "either that or came down here" I replied pointing to an open spot further ahead on the flat.

Bill stepped out of the bush and into the clearing. I was right on his heals, we looked up together as the ridge that partially shielded our view of the slope opened up as we strode across the clearing. Right in the middle of the slope, about two hundred yards away stood the bear, he was casually walking along a blow down heading for the very ground that we stood on. His cinnamon coat shone in the late afternoon's sun, as if on que and only as wild animals can, he somehow detected trouble, stopping mid stride. Bill and I had stopped at the same time and were down wind, he couldn't have heard us nor could he have winded us, it's uncanny how animals pick up that subtle nuance in their environment that tells them something isn't quite right.

The bear swapped directions on the log and looked away from us, upwind, his massive nose wrinkling as huge volumes of air were filtered through it. I took the opportunity and sat down, Bill stood as a statue to my side. The grizzly saw the movement I made sitting down and swapped ends again, looking in our direction.

"He spotted us" Bill whispered from the corner of his mouth.

"I got him", I said, peering through my rifles scope, "step back".

Bill slowly eased behind me, the bear was getting nervous, moving his huge head from side to side, nose held high, testing the airwaves. A raven croaked from high off somewhere, it's cries crystal clear in the cold mountain air, the bear took a couple steps along the log, "better take him" Bill said " he's going to take off".

I had come on this trip to find not just any bear but had hoped for one of the giants amongst giants that the area is known to hold, here before me in this grand place stood probably the second biggest grizzly I had ever laid eyes on. The biggest Bill and I had seen a couple years earlier a few miles north from where we stood. I had come, I had found, and now I had the shakes like never before. The bear took a couple more steps along the log and then stopped for one more look at what had invaded its domain, its massive head waving side to side as it tried to find answers in the airwaves.

That few second delay was enough, I managed to gather myself and settled the crosshairs, 178 grains of 7mm projectile reached out and found a home, the great bear fell off the log. We didn't run up nor did we walk up to the big bear, we waited an unbearably long time watching with our binoculars to make sure it was safe. After a while we climbed higher and came down on him from above.

The bear was huge; we estimated the boar to be somewhere around 650-700 lbs. It was more than I could have asked for, a huge boar, his feet still covered with the winter growth to protect them from the long cold winter buried beneath the snow, his 7' 10" cinnamon hide thick and unmarred. We both sat and marveled at the remarkable animal and the equally remarkable adventure we had in pursuit of it. I had pulled the trigger on this day but the bear was a success equally shared with Bill, neither one of us would have had the chance to harvest such an awesome animal without the other.

It was an arduous task skinning the immense bear on the hillside, and another task and a half to get the hide back to where our quads sat on the hillside a few miles away. We were both done in by the time we got back to camp that night and slunked our cold tired bodies into bed. We looked for a bear for Bill for another five days but didn't find another grizzly, we saw lots of black bears working the lower slopes and green spots but no more of the big bruins that we came for. Somehow it didn't really matter, we knew coming up that to get one bear would be great.

The trip out to the pick-up was less adventurous than the trip in but still held out for some interesting moments. The ten days of sun had melted most of the higher section so Bill rode the skidoo pulling the skimmer while I rigged up a rope harness and rode my quad and pulled his. We had to "skiquad" in a few sections along the roadbed but all in all the trip out was fine. The snow finally succumbed to the suns warmth about five miles from our pickup, Bill went ahead on his quad returning with the skidoo trailer hooked behind. We loaded the skidoo and skimmer on the trailer and made our way out to the pick-up. I'm sure if anyone saw the convoy coming down the road they would have scratched their heads trying to figure us out.

Mark Twain didn't want to have to explain away a typewriter "gadget" and we sure would've had a time explaining away the "weird-fangled" outfit we were herding down the old road!

The Mighty Moose Hunter

A Moose Hunting Adventure

*"With his rump to a stump, feet to a fire,
he dreams of a moose with antlers like a liar.
Dreaming while hunting is a very bad habit, you
usually end up with a snowshoe rabbit "*

How fitting I thought as that poem wiggled through my mirky mind. I was sitting against a stump, and wished for the warmth from a fire. And like the poem, my mind was drifting as my body wracked again and again with uncontrollable shivers. Thane Davies, a long time hunting partner, and I had both come down with a flu bug during the drive up. We had left Victoria, B.C. feeling fine. Sixteen hours later we arrived in north central BC feeling pretty rough. The weather that greeted us was damp and cold, not exactly conducive for sick people to be out and about in. Even though we were feeling the effects of the flu, a bull moose on the second day hunting raised our faltering spirits. Thane dispatched it with one shot from his 7mm. We were now looking for a cow moose.

British Columbia has a Limited Entry Hunt (LEH) lottery system for cow moose. While some management units (MU's) have open seasons, others require a LEH. The area we chose to hunt had both, immature bulls and calf's were open, mature bulls and cows required a LEH tag. Thane had applied for a bull moose and I for a cow. We both were lucky to receive LEH tags in the mail. My uncle Denis and his partner, Ray, had LEH tags as well, both of theirs being for cows. We had met them at our camping spot two days earlier.

Dark gray skies ominously hovered as I sat my vigil by the stump, another shiver wracked my cold body. I was wishing that I had stayed by the fire. Thane was coming up through the timber off to my left while I had climbed up and was overlooking a stand of red willows. Somewhere in the stand of trees was a moose. We had first found fresh tracks leading from a swamp into the trees yesterday morning, and again this morning a fresh set of tracks told the tale of a moose going from its feeding area to its bedding one. We were hoping that it would move up through the willows if disturbed. Yesterday I had come up through the trees and Thane had climbed to overlook the willow patch. Somehow the moose had eluded us yesterday but we were confident that today would prove different. I was lost in a daydream of warm fires when a loud "crack" echoed in the forest. Must be Thane, I thought.

Looking towards the timbers edge I watched a raven swoop down and land in a gnarly old aspen tree. The black bird was looking into deeper the forest. I watched it as it turned its attention to whatever was making the racket. Its head followed the unseen movement below. More cracks and rustling sounds hung like sentries in the misty cold air. The bird swapped ends on the branch as whatever was moving under its perch moved along. Seconds seemed like hours, the birds bobbing head followed the noisy progress, I followed the birds point. A dip in the terrain hid whatever was coming across from my view. Another loud sound came from something breaking a branch, then another echoed closer, but this one came from further down on the hill. "Hmmm", I mouthed.

Looking downhill at the area where the second noise came from I saw Thane standing there. Looking back at the raven I noticed that whatever was moving was soon going to crest the slope, its point was moving away from the tree. I shifted down behind the stump and I waited. The bobbing head of a moose came into view. "cow", my heart thumped.

I waited a moment to see if it was accompanied by a calf, it wasn't. The cow moved down the slope in earth eating strides. Its trail was angling towards a stand of thick willows. Moose are funny looking critters, most would agree the word "gangly" comes to mind when you see them. But those long legs equal long strides, when startled they can run at a remarkable pace. Resting my rifle across the stump I waited for an opening in the willows, the moose slowed to a walk. The cow stopped to look around but the willows prevented a shot. Carefully it moved forward until hesitating for a second in an opening. A loud "boom" shook the serene setting. The moose wavered at the shot, another shot and it was down. My trusty .270 loaded with 140 grain nosler ballistic tips had done the job.

A good rifle is a hunter's best friend, whether new or old, it is the tool that we always carry, seldom use, and performs when asked. Mine is old, my dad bought it new when I was nine. I have dragged it across tundra slopes, banged it on backpack trips, dropped it off cliffs, in water, in snow. Many years ago I decided to camouflage it. My wife said a new stock wasn't budgeted for, diapers were. I argued like a seasoned stockbroker but to no avail, diapers won out, I settled on painting the stock camo, glass beading the barrel and painting the barrel and scope with satin barbecue paint. It isn't as pretty as a new camo stock, nor as attractive as a stainless rifle but year in year out it gets the job done.

My big wet sniffly grin met Thane as he came over the hill top. We "all righted" "yehaad" and back-slapped each other. Thane and I have shared many such moments. I had pulled the trigger on this moose, but he had shook it from its hidy hole, we both were successful. The cow looked to be about three years old, she would fill the freezer with many pounds of mouth-watering meat. Thane was suffering from the same flu symptoms as I. We lit a fire and warmed ourselves with a nice hot coffee before the arduous task of dressing and packing the moose out. We had a half mile pack to reach the road. Cutting the moose into eight pieces we set about the tough job of moving the moose to the road. My uncle and Ray came along just as we lugged the last two pieces down the hill.

"Good timing Eh" Denis said.

Puffing hard Thane said that they had probably hid around the corner until we retrieved the last pieces. The dig was met with a good natured laugh. We tied all the pieces on our motorcycles and headed back to camp.

We still had a week of holiday time left but were tagged out. Over dinner that night Denis and Ray asked us to stay and help them find them a moose or two, they hadn't had any luck yet. It didn't take much convincing, we were having such a good time. Thane and I would carry shotguns and look for grouse, while accompanying them as they looked for moose. The week passed way to fast as most good trips do. Neither of the other two hunters connected on moose. And I'm not sure they minded either, we enjoyed sharing camp with them and had good times searching for moose. In retrospect I was glad we stayed. Thane and I had hunted with my uncle for years, shared miles of trails, many successful hunts and many not successful. We had a mutual love for being in the wilderness. Denis taught Thane and I to focus on the adventure, opposed to the kill, to cherish and respect the privilege of hunting and wildlife. I think that his tutelage enabled us to become better hunters. Sadly and unexpectedly Denis passed away the spring after that hunt.

Many moose have fallen to my gun since that hunt, some huge big old bulls with sheet of plywood size horns and some freezer filling cows. But none match the memories from that fateful hunt years ago, it remains fixed in my brain as what hunting really is.

A hunt comprises of many things, the kill is but a single piece in the whole event. A good hunt is companionship, planning, sharing camp fires, trekking tough trails, rough weather, laughs, hard work, anguish and more. Sharing a trip, day, kill, with good people that's the meaning of hunting not just the game taken. A hunt can be just as successful without squeezing the trigger. Too many hunters feel that a trip without harvesting their quarry is unsuccessful. Primitive hunters didn't care about the headgear of their prey, modern day hunters stand transfixed on them. I have had the good fortune to harvest many animals, some possessed awesome headgear, huge bull moose, bull caribou, stone sheep rams, mountain goats and a number of incredible mule deer. Those trophies pales in comparison to a four point black-tail deer I took and shared with Thane and Denis. We had relived that hunt over and over and they both congratulated me on the trophy. I would gladly trade the "trophy" of a good adventure, for a physical "trophy" without adventure any day.

Good partners are hard to find, Denis was more than a partner. Through his understanding of what hunting really is all about, Thane and I got "it"...

Of a Quest for Bulls

A Hunting Adventure

*"Fortitude is the Marshal of Thought,
Armor of the Will and the Fort of Reason"*

The famous philosopher Sir Francis Bacon penned that quote in the late fifteen century, far before me or my partners Bill Cash or Geoff Helfrich were born. However, at the moment, as I sat nestled under a tree on the side of a ridge tracking an enormous bull moose in my binoculars, I thought that perhaps Bill could have been the model for that maxim. "That is a great bull" Geoff whispered in the growing wind, sharp crisp snowflakes came from the south stinging my cold cheeks.

Bill, Geoff and I were camped on the south side of a small hill in north central BC. Tethered or hobbled in the grassy meadow below our camp were the horses we had ridden and trailed into the mountains. Spread out before us, as far as you could see, were mountain tops and high alpine valleys. It is a grand place. This was bull moose and bull caribou country and we had all been there many times before. It was why we had traveled so far to get to, although it was with some luck and with a generous helping of fortitudinous on Bills part that we even got there.

But I am getting a bit ahead of myself.

Four days earlier the three of us had loaded six horses, all the gear needed along with two weeks-worth of "add water or cook over a fire" food into a pick-up and horse trailer. After a seventeen hour drive, twelve hours of that being gravel, we arrived at our starting point. We decided to spend the night camped by the truck before heading up a river that we would follow through a mountain pass into the plateau beyond. By dark that first night we had watered the horses, fed and tethered them and slipped into our sleeping bags. The next day was going to be a long one.

The next morning greeted us in typical early September high mountain fashion: a brisk wind blew from the north, the skies were blue as far as we could see, surrounding mountain tops were covered with a fresh dusting of snow, all the colors of fall were prevalent in the river basin and valley bottoms. It was going to an awesome day for a ride. Geoff and I led the horses down to the creek for a drink while Bill set about getting a fire going. We were going to have a quick coffee and saddle up for the thirty mile ride to a line cabin further up the valley.

Spirits were high as we loaded and tacked our string. Just as Bill stepped up on the horse he was riding it bolted with him half in the saddle. He held on as the horse gained speed down the roadway. I was just thinking he was going to get the horse stopped when it juked right and he didn't.

"Oh crap" Geoff yelled as Bill flew head first into a pile of rocks.

I jumped off my horse and ran the hundred and fifty yards to where he was piled up.

Bills horse, Diesel, had continued his frantic escape and was another hundred or so yards down the road when I got to Bill. He was out cold and bleeding profusely from a huge wound on his forehead. I gently rolled him on his side as he came to.

"Bill, Bill, whoa buddy just stay still for a minute" I said

Bill was slowly coming too and had that "oh shit" frantic look in his eye. He rolled onto his side and was reaching for his forehead, it was a bloody mess. Geoff came up as Bill gingerly sat up, his hip hurt from landing on a rock however the real worry was his head. He had landed face first in a pile of rocks and had really rung his bell, there was a big cut above his eyebrow and a pretty good size egg on his cheek bone.

The cut was bleeding badly but as I washed it with cold creek water and paper towel the bleeding slowed. We were in a dilemma, help was twelve hours of gravel road away and even if we went I wasn't sure much could be done. Bill showed his mettle and said, "Let's just hang out here for a couple hours and see how things go".

While Bill sat and recouped Geoff and I decided we would untack the horses and stay there, the possibility of heading back to civilization and medical help was real and we wanted to be ready to get on the road if we needed to. But, after a few hours sitting in the morning sun, Bill got his feet under him and was moving about, he definitely had a concussion and was going to be sore but we thought we would be okay staying.

As the day progressed Bills head got clearer but his hip got worse. We decided to stay where we were for a couple days and let him heal up. Two day later we again tacked up the horses and headed for the distant cabin. Aside from a couple stops along the way to tighten saddles the day was just a glorious fall ride through a mountain pass. Late in the afternoon we had gotten to the cabin and were sitting around a campfire.

The next morning brought another magnificent day to be riding in the mountains. The sun shone warm, dickie birds cavorted along the creeks. Before long we were up on the plateau and heading for our distant camp. By nightfall we had camp set up and horses grazing.

"Well, a bit late but we are here" Bill said "You are a tough cookie" Geoff said to Bill

Just as light was leaving the valley top we sat around a small fire and made plans for the next day. Geoff and I were going to hike to a ridge where we would be able to glass the entire valley bottom below, Bill wanted to circle around and would meet us up there later in the day.

"That's a dandy bull" Geoff whispered

We were at our "one" spot and had our spotting scopes zoomed in on a big bull moose that was with five cows down below us by a small creek. He was milling about moving from cow to cow, presumably checking them out. Off to our left a mile or so away a smaller bull stood on the edge of a copse of trees absorbing the morning sun's rays.

"Let's watch and see where they go" I said not taking my eyes off the bull.

I was just trying to decide the best course to intersect them when the lead cow took off upstream away from us. As we watched they crossed the valley floor and disappeared into a stand of trees a couple miles away.

"They will be back" I said

Geoff and I shouldered our packs and moved further along the hillside. At our "second" spotting place we sat down and Geoff let out a few soft cow calls, waited a few minutes and let out a few louder calls. I was scanning the valley with my binoculars as he called, almost immediately I heard a distinctive "Urph" from the trees below.

He cow called again...

The bull answered again...

I couldn't see the bull moose but we could sort out where it was by his calls. As Geoff called the bull answered and we could tell it was moving. Before long he stepped out on the valley floor below us, it was a huge bull with red horns. Apparently he had just rubbed the velvet off his rack.

"Whoa" I said, "that thing is huge"

As we watched, the bull started making his way towards where we were hidden, weaving in and out of stands of trees as it came. Geoff kept calling but quieter and quieter as it got closer.

"It's going to come up that drainage", I said pointing to a small ravine that ran down the mountain side. "ya probably" Geoff replied.

I had just saw the red bull when another bull answered from below us. We looked at each other with big grins.

The second bull was coming as well, minutes later we saw him about seventy five yards below. He was a young bull nowhere near the size of the red bull. Geoff quit calling but the young bull kept coming, in no time he was twenty yards below us full broadside and looking for the cow. We froze in place, the young bull kept coming. At about ten yards he caught wind of us and bolted, straight to where the red bull was!

"Dam" Geoff said

We watched as the little bull crashed away down the ravine. As if on que we saw the red horn bull come out of a stand of trees and take off downhill and out of sight.

"Crap" I said "oh well let's just stay here for a bit and see what happens, that big bull isn't going anywhere"

We were sitting there letting everything settle down when Bill snuck over the ridge and joined us. He had circled the base of the mountain and came up but hadn't seen anything. The three of us sat there the rest of the day calling and watching, as dusk fell on another great day in the mountains we hiked back to camp.

Over coffee the next morning we decide to go to spotting spot "one" for a bit and Bill was going to go up valley and meet us in the "second" spot later in the day. Geoff and I had just got to our spot when I spotted, way up the valley a terrific bull caribou trotting along the creek.

"Geoff" I said "there's your bull"

He dropped his pack and found the animal in his binoculars. "Wow" he said "that's a good one"

"Yup" I replied "let's head down to that high spot in the valley floor, I think that caribou is going to come right down the creek.

We grabbed our back packs and headed down hill to the spot. As the sun broke over the ridgeline we crawled up on the hump and looked for the bull. He was nowhere to be seen but that wasn't a surprise, the valley bottom had lots of humps and ravines. The wind was in our favor so we decided to stay put and let the bull come to us. About an hour later we saw him, he was in the creek bed about five hundred yards upstream, instead of moving our way though it looked like he was working up the opposite hillside.

"Let's go" I said to Geoff

"I think we should cross the valley floor and sneak along the other side" he said, "good plan" I replied. On the other side of the creek bed we found an animal trail forty yards up the bank from the creek so we quickly closed the distance. Before long we were just across a ravine from the bull. I ranged him "one hundred and forty yards" I whispered to Geoff as he laid his rifle over his pack. The bull was milling about in a stand of willows eating and beating the bushes with his tall horns.

"Get ready" I said as the bull stepped in the clear.

Just as the caribou was clear Geoff shot, the bullet from his .300 magnum hit its mark and the big bull dropped in his tracks.

Geoff was smiling from ear to ear as he walked up to the bull, it was huge caribou, bright white mane with very tall and wide horns.

"That is an awesome bull" I said to Geoff.

On a beautiful sunny fall afternoon, fifty five miles from the nearest road we sat and marveled about the great adventure the day offered. After taking a bunch of pictures we set about skinning and quartering Geoff's animal. Our plan was to quarter it, cut some brush to cover it to keep the crows off it and go get the horses.

"Let's take the cape over to the creek and wash it out on the way" I said, "bring the horns as well we can hang the cape over them when we are done"

We made a tee pee out of the quarters and covered them with brush and whatever else we could find and hiked over to the creek. Geoff rinsed the cape out and hung it from the horns in the wind.

We were about four hours from camp so we thought it might be the morning before we got back with the horses "it'll be fine" I said and started back the way we came.

At one point just before we got up to the ridgeline we stopped for a break and scanned the valley floor for animals. Looking up stream I saw a cow moose standing on the edge of a little clearing "that's weird" I said, "one cow all by herself this time of the year". "There must be more in the trees or laying down" Geoff said. No sooner had he said that a massive bull moose stepped out from the trees.

"Holy doodle" I said "that thing is a tank"

We didn't have time left in the day to go after it so we sat there and watched it. "Oh man, I hope he is there tomorrow" I said.

We hiked into camp an hour after dark had claimed the high country. Bill had a fire going and a hot pot of coffee all brewed. Geoff pulled the back straps out of his backpack and before long we had them cut up and frying in a pan over the fire.

"What a great day guys" Bill said, "I saw a couple decent bull moose and a bull caribou but couldn't make out his horns". In between mouthfuls of fresh caribou meat we told the days stories around the campfire. It indeed had been a great day.

The next morning we tacked up the horses for the ride over to retrieve Geoff's bull. Stopping at the second spot we tried to find the big bull moose we had seen the day before.

"There he is" Bill said, "up higher in the brush" just as he said it I found him in my binoculars "what a dandy" I said.

"Okay you guys take the horses down the next drainage", I said "I am going to side hill over a few more and see if I can get that bull".

"Okay" they both replied "good luck"

The wind was gently breezing in my face and the sun had yet to crest the ridge, it was cool and clear. I figured I could just zip along the tree line and then drop over a ridge right into the little meadow the bull was in.

Just before I went into the meadow I stopped and loaded my rifle. Sneaking from tree to tree, then to the last tree I quietly got to where I thought I was one hundred yards or so from him. I had no sooner stopped behind the last spindly little spruce tree when a cow moose stood up about seventy five yards from me, she had me pegged. I froze in place. Then another cow stood up, then another, then a calf stood up and looked right at me. I was barely breathing. One of the cows started looking around in that frantic something isn't right way. I didn't know what was going to happen or where the bull was but I was standing there, quiet as a mouse, still as a statue when the bull stood up fifty yards away to the right of the cows. He stretched, swung his enormous rack around and took a step towards the cows. The cows all looked in his direction, I raised my rifle found the crosshairs and shot.

The big bull didn't take a step, he just fell over. The cows and calf bolted for distant meadows. I waited a few minute then took a big circle to come down on the bull but it didn't matter, he was dead. And he was way bigger than I thought, "what a tank" I said to myself.

I snapped a bunch of pictures and had just started the arduous task of skinning and quartering the bull when I heard whistles from up on the ridge. Geoff and Bill had heard the shot and came to help me, quartering a bull moose is a big job so I was glad for the help.

We spent the day hauling the moose up to the ridgeline, stashing it and then went and got Geoff's caribou, stashing it in the same spot. We would come back and load for the long trip out to the truck from there.

As night was shouldering its way into the valley we crossed the ridge towards camp astride our horses. About an hour from camp I spotted a moose out on the valley floor in willow patch, he was thrashing the bushes with his horns. We got off our horses and sat down to have a better look, it was the red horned bull and he was in a spot where we could get at him pretty easy. We whispered back and forth but with two animals down and miles to get back to the pickup we just didn't have room for another animal.

"We will get him next year" Bill said "and he will be bigger"

I could only look over and smile at Bill, he certainly exuded the elements Sir Francis Bacon penned in his quote. The next morning we took camp apart, went and loaded the caribou and moose quarters onto our horses and headed back to the truck. The trip out is a two day event so our plan was to get to the little cabin sitting on the shores of the river that night. About half way across the last plateau I spotted a caribou in a distant bowl.

"Look at that" I said to Bill and Geoff, we all got off our horses and looked at the bull through our binoculars.

It was a nice bull but it was way down in a bowl so we kept going. As we crossed an open ridge I looked again and saw that the bull was coming our way.

"Keep going" I said to Bill "Geoff and I will just wait here and see what it's up to, see you at the cabin".

A rain squall washed across the plateau as Geoff and I moved along, leading our rider and pack horses. Bill was slowly engulfed and disappeared in the increasing weather. Fog and rain squalls flowed across the plateau.

Geoff and I slowly kept going, at time we could see the bull other times he was gone. The tundra is a weird place, it looks flat as a pancake but in reality the terrain is rolling, animals can disappear without a trace. Just when we thought that the bull must have went away Geoff said,

"Dawson there he is"

Above us, on a small rise, the bull came over and stopped, full broadside and full silhouetted looking at us. There wasn't any need for binoculars, he easily passed the legal requirement. I sank down on the cold wet ground and got ready.

"Hold the horses" I whispered to Geoff

He grabbed all the lead ropes and stepped over a bit, the bull swapped ends and moved a few yards closer, staring at the horses.

"He's getting antsy" Geoff said

Just as he said that I squeezed the trigger. The bull stepped ahead a bit and fell.

"Great shot" Geoff said

The bull was magnificent, it was getting late in the day so we snapped a few pictures rolled him over, dressed him and left him to deal with until the morning.

"Let's take the loins for diner" Geoff said.

It was dark by the time we continued towards the cabin.

"It'll be fine overnight" I said, "the birds won't find him tonight so we just need to get back early in the morning"

Riding into camp a few hours later we were greeted by a nice roaring fire outside and the cabin had a fire in the little wood stove. Geoff and I were soaked to the bone but quickly changed to dry gear and told Bill about the last minute bull.

"We'll ride up early tomorrow and pack him out" I said to Bill.

It was raining in the am as we saddled our horses and started up to the plateau. As the caribou came into view we could see that something had found it in the night

"Crap" Geoff said

It appeared that a fox or maybe a lone wolf had found the bull in the night, one hind was chewed up pretty good and the ribs had been gnawed on. We assumed whatever it was must have seen us coming across the plateau and took off.

"Oh well" I said "there's not much we can do about it"

As we quartered the bull Bill loaded it on the pack horses and before long we had it all looked after and were heading back to the cabin. We spent another night at the cabin, enjoying the warm fire and cooking fresh caribou.

The next morning we got up early, loaded the horses up and started the thirty mile ride back to the pick-up. A brisk wind blew from the north, skies were blue as far as we could see, the surrounding mountain tops were covered with a fresh dusting of snow. All the colors of fall were prevalent in the river basins and valley bottoms. It was an awesome day for a ride, as we rode along I reflect on the previous two weeks, what an adventure we had…

About the Author

Dawson Smith has been hunting for over fifty years and has hunted all over the world. Throughout his travels, Smith has found British Columbia to be the go-to location for the prime hunting experience. And he's learned that the perfect hunt is achieved through a balance of the spirit of adventure, luck and risk. *Tales from the Trail* is Smith's first collection, detailing such experiences.

Smith lives on a horse farm in Prince George, British Columbia with his partner, Sandra; five horses; and three Great Danes. For more of Smith's works, see publications such as *Big Game Adventures*, *Western Sportsman*, *BC Outdoors*, *Big Buck Magazine*, and *Successful Hunter*, in which other of his short stories have been published over the years.

bndproductionsca@gmail.com
www.bndproductions.ca

CPSIA information can be obtained
at www.ICGtesting.com
Printed in the USA
BVHW060730041020
590213BV00002B/2